第一卷
创意面料设计
Creative Fabrics Design 1

沈 沉 著

唐再凤　译
范秀华

大连理工大学出版社

图书在版编目 (CIP) 数据

创意面料设计：汉英对照 / 沈沉著；唐再风, 范
秀华译. —大连：大连理工大学出版社, 2013.6
ISBN 978-7-5611-7854-6

Ⅰ. ①创… Ⅱ. ①沈… ②唐… ③范… Ⅲ. ①服装面料 – 设计 – 汉、英 Ⅳ. ① TS941.4

中国版本图书馆 CIP 数据核字 (2013) 第 105156 号

出版发行：大连理工大学出版社
　　　　　（地址：大连市软件园路 80 号 邮编：116023)
印　　刷：利丰雅高印刷（深圳）有限公司
幅面尺寸：230mm×300mm
印　　张：20
插　　页：4
出版时间：2013 年 6 月第 1 版
印刷时间：2013 年 6 月第 1 次印刷
责任编辑：裘美倩
责任校对：王秀媛
封面设计：沈　沉

ISBN 978-7-5611-7854-6
定　　价：298.00 元

电　话：0411-84708842
传　真：0411-84701466
邮　购：0411-84703636
E-mail：designbook@yahoo.cn
URL：http:// www.dutp.cn

如有质量问题请联系出版中心：（0411）84709043　84709246

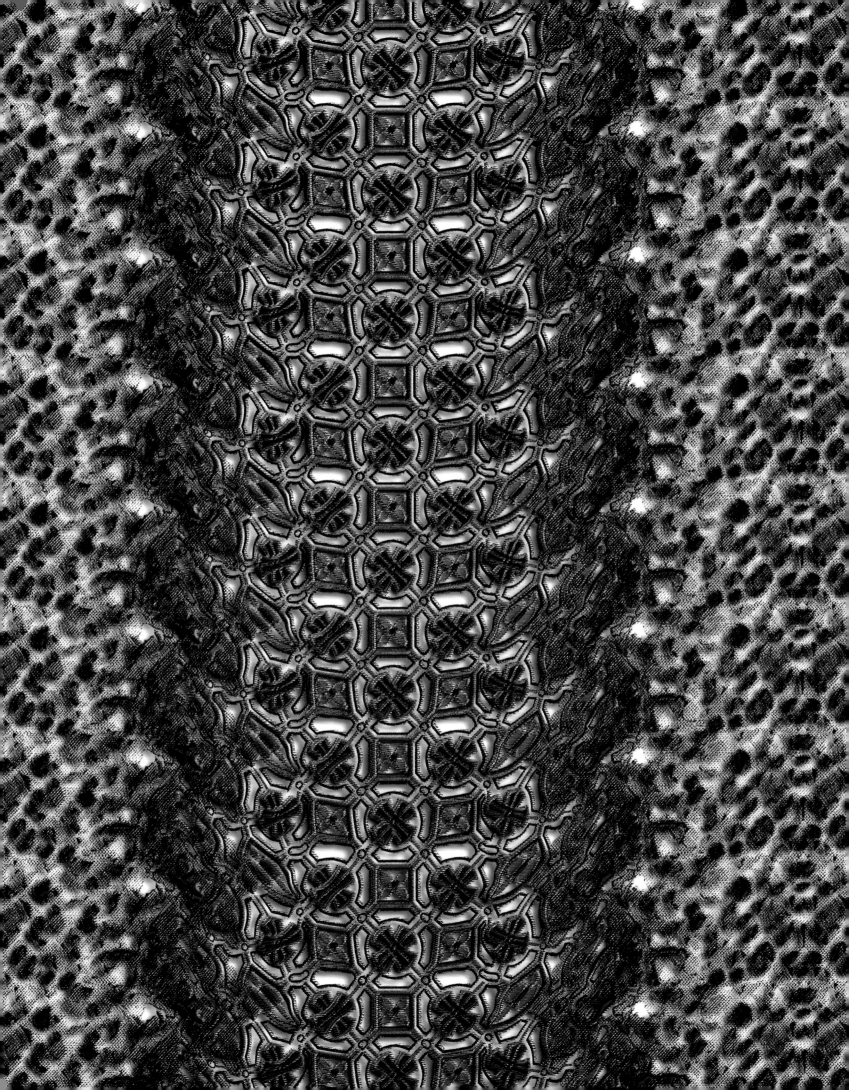

业之真

　　品读沈沉老师的《创意面料设计》，是一次惊艳的视觉之旅。

　　自然、科技、生活、民俗……一个个美的瞬间构筑了沈沉老师纺织品创作的灵感之"道"，吸纳天地精华，传承多元文化，娴熟的数码技艺在"法""理"之间跌宕起伏，依托纺织创新技"术"，凭借时尚流行趋"势"，在源自不同历史背景和蕴藏不同文化内涵的审美观与价值观的交融、凝聚、升华之中，经典元素和手工技艺被重新审视和弘扬，并融入了新时代的时尚格调，一幅幅创意面料演绎着色彩、肌理、组织、纹样彼此交互的新奇视觉和曼妙风情，带来了超现实主义出其不意的美感，设计的生命力因此而充盈丰沛、绚丽怒放。

　　品读沈沉老师的《创意面料设计》，也是一次欣喜的希望之旅。

　　基于坚实的产业规模、完整的产业链配套、先进的技术装备等综合比较优势，中国纺织工业具有良好的市场竞争力；然而，随着市场容量增长有限，生产成本持续增加以及经营风险加剧等竞争环境的变化，产品结构、市场结构、商业模式的调整已成为必然，建立创新机制与培养创新人才成为纺织强国的关键要素，对于创意设计的期待成为全行业关注的话题。沈沉老师运用面料设计语言表达了对提升纺织产品附加值的思考，并将这种思考与创意实践相结合，通过作品中喷薄而出的原创激情与视觉力量，实现了纺织生产技术在文化创意和精神内涵方面的价值延伸，着实令人欣喜和骄傲。

　　品读沈沉老师的《创意面料设计》，更是一次深刻的反思之旅。

　　中国时尚产业反映了我们所处的时代特色，一方面在文化传承中求新求变；一方面在学习中寻找符合自身发展的道路。人口数量增长及结构变化赋予消费市场多元、多变、复杂、包容的特征，产品生命周期更为短暂，消费者更加趋于理性，消费诉求在产品市场价格维度和品牌心理价值维度之间寻找平衡，对于创意的需求更趋向于独特的感觉、新颖的概念或精神的认同。成熟的设计既要注重物质消费形态设计和感性消费形态设计，更需体现符合生态文明的社会责任和对人性光辉的崇敬，必将更多地建立在思想和创意之上而非依靠复杂的工艺与大量的能耗，必将更贴近人们的心灵和生活，从生活中来，回到生活中去……展望未来，中国需要自主的科学技术、自主的创意设计与自主的时尚品牌，创意经济必将从规模经济中脱颖而出，中国将不仅是世界最大的加工厂，更是世界的时尚策源地，这便是当代中国设计师群体的内在驱动力。

　　深切期待沈沉老师在教学中为中国纺织行业发展培养出更多富有原创精神和责任意识的纺织面料设计新秀，在设计实践中诞生更多、更精彩的设计佳作，复兴中国丝绸之路，成就人类美丽生活。

国家纺织产品开发中心主任　李斌红

2013年3月9日

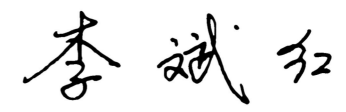

The Truth of Profession

Reading Mr. Chen Shen's Creative Fabric Design is a breathtaking journey of visual beauty.

The way Mr. Shen is inspired in his creative works of textile has come up with so many beautiful moments out of, respectively, nature, science and technology, living, and folk customs, each of which has assimilated the essence under heaven and inherited multi-cultures. He used sophisticated digital techniques in a way in which method and reason were quirkily touched. With the aid of textile technological changes while keeping up with fashion trend, he invigorated classical elements and traditional craftsmanship by blending them with today's fashion style. He achieved this through a process that values and interprets different historical backgrounds and different cultural connotations, integrating them and agglomerating, leading to sublimation. One by one, his original fabrics illustrate newfangled vision and elegancy, which are brought about by the interactions among color, fabric texture and pattern. As a result, a surrealist fantasy of beauty beyond one's expectation has been produced; the talent of a designer is as in full blossom.

Reading this book is also a happy journey of hope.

China's textile industry enjoys its competitiveness in the market, owing to its comprehensive comparative advantages, including its large scale of production, complete industrial chain, and advanced technology and equipment. Nevertheless, as the market is saturated and the production costs and business risks are ever growing, it is of necessity that the production structure, market structure and business model will be changed. The key to establish a leading textile power is to introduce mechanisms of innovation and cultivate innovative talents. Creative design has become a topic of concern throughout the industry. With the language of fabric design, Mr. Shen has expressed his thinking about how to increase the added-value of textile products and practiced his thinking as well. Through the original passion and vision power rushing forth from his products, he has extended the value of textile production technology into cultural innovation and spiritual connotation. It is worthy of joy and pride.

Reading this book is further a journey of profound introspection.

China's fashion industry reflects the characteristics of the times in which we live, i.e. to pursue novelty and change along with cultural inheritance on the one hand, and to find a way suitable for our specific stage of development through learning on the other. Population growth and demographic changes make the consumer market more diversified, quick-changing, complicated and inclusive. Product life cycle has shortened. Consumers have become more rational, demanding a balance between the market price of the product and the psychological value of its brand. The demand for creativity tends to be more of a unique feeling, novel concept or spirit identity. In addition to material consumption and perceptual consumption, a mature design should also pay more attention to the social responsibility required by eco-friendliness as well as pay respect to human nature. Design will be based more on thoughts and ideas rather than on complex processes and huge amounts of energy consumption. Design will be closer to people's soul and life, and serve living by learning from living...Looking forward to the future, China needs its own science and technology, its own creative designs, and its own fashion brands. Creative economy will certainly break out from the economies of scale. China will not only be the world's largest manufacturing plant, but also a source of the world's fashion, which is the internal driving force of the designers group in current China.

I truly hope that Mr. Shen will train more fabric design talents with originality and responsibility for the Chinese textile industry in his education career and that he will further create more wonderful fabric design works in his practice of fabric design, so as to make contribution to the revival of China's Silk Road and the course of creating a beautiful life for human beings.

Binhong Li , director of the National Textile Product Development Center
9th March, 2013

善之学

  美丽中国是每一个中国人的梦想，更是中国设计教育工作者的责任。"中国创造"的建立，要求设计教育坚持创新思维、勇于探索实践，要求有一批有知识更有见识、有传承更有创意、有使命感的教育工作者和创造者。

  东华大学作为国内首批创立纺织服装类学科的"211 工程"国家重点高校，有一批倾心教育事业、教学科研杰出、坚持创意为先的艺术设计教师，服装·艺术设计学院纺织品艺术设计专业教研室主任沈沉老师就是其中之一。沈老师在教学与专业设计上始终坚持原创、坚持产学研的结合，多年来保持着每年八九百件原创作品的面世。本书中大部分设计作品是沈老师2012年下半年以来的创作，其灵感来源于其国内外实地拍摄的摄影作品。沈老师的纺织品数字艺术设计在种类、工艺和表现手段上体现了"必先创异、创造不同"的理念，其设计鼓励异化，追求异端。同时，他追求设计体现创益，即创造"公益、效益、利益"，其作品不仅在教学中展现了强大的示范效应，也给产业界提供了珍贵的比较和选择，产生了良好的社会效益和可持续发展经济利益。

  学生心目中可敬可爱的沈老师，有一个更可贵之处就是坚持与同学们一起做作业、一起参赛、一起做市场调研、一起做科研……与同学们一起分享设计带来的愉悦，一起分享设计带来的成功和挫折。他支持同学们参加大赛，将学生作品放到更广的评价体系中检阅；他坚持理论联系实际，鼓励学生作品走入市场；他极力倡导并身体力行带领同学们到中国各地原住民及文化地区考察，探究本土思想下的符合国际潮流的原创设计。

  中国纺织业正处于从纺织大国走向纺织强国的转变，创新是关键所在，东华大学正因为有很多像沈老师一样热爱教育事业、专业能力强，思想独特的老师坚持不懈地努力和探索，才成为中国纺织行业向世界发出中国创造新思想、新力量、新标准的地方。我们一起努力！

东华大学副校长
服装·艺术设计学院院长
刘春红博士
2013年3月17日

The Learning of Goodness

Beautifying China is the dream for every Chinese, and it is more of a responsibility for the Chinese design educators. To establish "Originated in China" requires that design education should persist in innovative thinking and bold exploration and practice. It also requires a number of educators and designers, who not only are knowledgeable but also insightful, not only inherit the tradition but also are innovative and have a sense of mission.

Donghua University is the first of the national key universities of the "211 Project" to found textile and fashion disciplines. Working with the university there is a group of art and design teachers who love educational career, have achieved highly in teaching activities and academic research, and have always given innovation a top priority. Among them is Mr. Shen Chen, head of the section of Textile Design under "Fashion and Art Design Institute" of Donghua University. Mr. Shen always endeavors to innovate and insists on combining teaching with research and production. Over the years, he has been publishing 800 to 900 pieces of original design works annually. Most designs in this book have been created by Mr. Shen over the last several months, the inspiration of which was from his photographic works of field shooting at home and abroad. Mr. Shen's textile digital art design embodies the idea of "the first to create, and create differences" in types, procedure, and means of expression. In his design he encourages "alienation" and seeks "heresies". At the same time, he pursues designs that bring about public welfare, benefit, and profit. His works not only show strong demonstrative effects in teaching, but also provide rare and precious possibilities of comparison and choices for the industry, thus serving well both the social welfare and sustainable economic development.

Respected and loved by his students, Mr. Shen is especially appreciated that he insists on joining the students: helping with their assignments, participating in contests, doing market survey, and carrying out scientific research, etc. He shares the pleasure and tastes sweetness of success and bitterness of setbacks all along with his students. He supports his students to participate in competitions, so that their works of design are able to be judged in a broader evaluation system. He insists on integrating theory with practice, and encourages his students' works to be marketed. He strongly advocates and sets an example by personally taking part in leading the students to travel all over China to the aboriginal and cultural areas to study, exploring original designs of local ideology in line with the international trend.

China textile industry is currently in transition from a big one to a powerful one. Along with that metamorphose, innovation is the key. Donghua University has become the place in China to radiate new ideas, new power, and new standards of textile industry to the world, precisely because it boasts many teachers with creative minds and expertise and dedication to education like Mr. Shen who have been endeavoring and exploring perseveringly. Let's work together!

Dr. Chunhong Liu, Dean of the Institute of Fashion&Art Desion
Vice-president of Donghua University
17th March, 2013

美之研

  作为一位本土服装设计师，对于中国自主创造开发面料的渴望一直在我内心深处。中国一直是纺织服装制造大国，在今天大环境的变迁之下，产业升级与自主创新绝对是当务之急。而对当前国内纺织服装产业而言，我们缺乏的是一个能告诉大家方向与走向的权威人物或是机构，能整合时势，深入人心，从历史层面及文化层面来体会人们的感受，再结合国际大环境，推出代表中国、属于中国的流行趋势，触动人心，引起共鸣。

  我很欣喜地看到沈老师《创意面料设计》一书的出版，这是我们中国设计师的一盏明灯。作为中国时尚产业的一分子，如果我们能够不再依赖国外面料，那将是我们的光荣，我们才不愧被称为中国设计师。而面料的设计研发，则需要一位像沈老师这样的领军人物。

  再进一步想想，若是整个时尚及流行产业都能如此，大家都以此为方向来做设计研发，虽然各自领域不同，但是异曲同工，最后出来的作品都能相互辉映、自成系统，带有强烈的中国文化特色，其中所蕴含的文化与精神，绝对能让国际人士刮目相看，引领时尚、创造潮流。

  这是我们中国设计师们心之所向，意之所属。沈老师身负重任，而在这方面的成就必定不负众望。

郭培
于北京玫瑰坊
2013年3月18日

The Study of Beauty

As a local fashion designer, I have long been looking forward to the day that China can create and develop fabrics independently. Being a big textile and clothing manufacturing country for a long time, today's China has to undertake industrial upgrading and independent innovation, an imperative task raised by the changing environment. But at the moment, China's textiles and clothing industry is in need of some authority figures or institutions to guide directions and development. Such figures or institutions should be able to react to the times, so as to speak to the heart of people, by understanding people, discerning human feelings in historical and cultural context, and keeping in line with international trend at the same time. Then we have every right to hope that the fashion trends, which represent China and belong to China, heart-touching and resonant, will come into being.

I am glad to see the publication of Mr. Shen's Creative Fabric Design, which is a shining beacon for the Chinese fashion designers. As members of China's fashion industry, we take it as our glory that one day we no longer depend on foreign fabrics, and only then do we truly deserve the title of Chinese designers. Mr. Shen is exactly this kind of front runners we need in fabric design and development.

Let's consider it a step further. If this is recognized throughout the fashion industry and the design and development are carried out in line with this direction, we will finally reach our common goal in spite of the fact that we all have our diverse domain, working in different fields of the industry. All the final works thus produced could contrast with each other pleasantly but contribute to a unique system with Chinese touches of characteristics, and the culture and spirit embedded wherein will certainly impress people all over the world, therefore leading the fashion trend and creating fashion style.

Herein are what we Chinese designers have long hoped for and been motivated to as well. Mr. Shen is carrying the ball and his achievements in this respect are bound to be expected.

Pei Guo
Rose Studio
Beijing
18th March, 2013

目录
Contents

设计作品 Design Works	11
后跋 Postscript	316

设计作品
Design Works

Creative Fabrics Design 1

名：富兰克林蚕缀的种子水纹列（8）
道：道法自然
理：近似、交互渐变、方向性、表面革质
法：四方连缀、纬向接回21.6厘米
术：梭织提花、皮革丝网印、轧花、嵌条
势：2013/2014国际流行趋势之蛇皮等甲胄形态
器：服装面料
Name: Water Wave Column of Franklin Arenaria Seeds (8)
Conviction: imitation of nature
Principle: approximation, interactive gradient, directivity, leather surface texture
Method: 4 sides in continuation, zonal tying-in 21.6 cm
Operation: jacquard weaving, leather silk screen printing, embossing, banding
Trend: 2013/2014 international fashion trend of armor feature as of snakeskin
Application: clothing fabrics

时间：2011年2月8日
地点：坦桑尼亚塞伦盖迪国家公园
对象：蜥蜴
灵感：甲胄、光泽、渐变
Time: February 8, 2011
Location: Serengeti National Park, Tanzania
Object: the lizard
Inspiration: armor, luster, gradient

名：花园格栅镂空（5）
道：道法自然
理：近似、交互渐变、方向性、表面革质
法：四方连缀、纬向接回5.5厘米
术：梭织提花、皮革丝网印、轧花、嵌条
势：2013/2014国际流行趋势之蛇皮等甲胄形态
器：服装面料
Name: Garden Grille Hollowing (5)
Conviction: imitation of nature
Principle: approximation, interactive gradient, directivity, leather surface texture
Method: 4 sides in continuation, zonal tying-in 5.5 cm
Operation: jacquard weaving, leather silk screen printing, embossing, banding
Trend: 2013/2014 international fashion trend of armor feature as of snakeskin
Application: clothing fabrics

名：E趋势时代——骇客帝国之机器虫（2）、（3）
道：科技主义
理：镜像、节肢动物、金属质感、水果质感
法：四方连缀、经向接回64.2厘米
术：丝网印、轧花、涂层
势：2013/2014国际流行趋势之科技形态
器：装饰布、服装面料

Name: E-era Trend—Hacker Empire Machine Worm (2), (3)
Conviction: science and technology
Principle: mirror image, arthropod animal, metal texture, fruit quality
Method: 4 sides in continuation, zonal tying-in 64.2 cm
Operation: screen printing, embossing, coating
Trend: 2013/2014 international fashion trend in science and technology
Application: decorative cloth, clothing fabric

Creative Fabrics Design 1

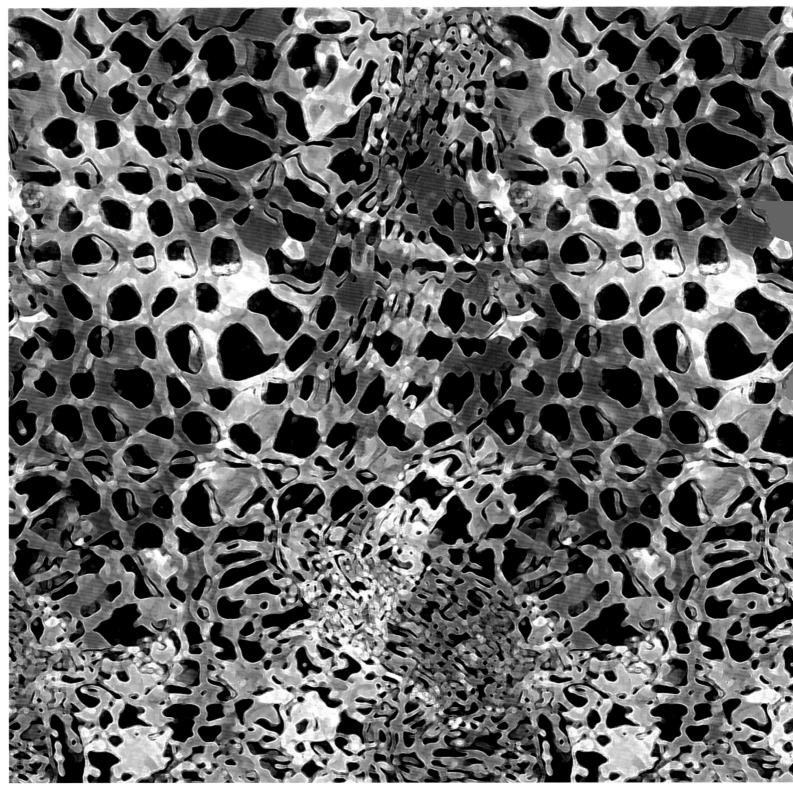

名：豹纹印象（3）
道：中国祈福文化
理：近似、交互渐变、金属质感
法：四方连缀、经向接回64.2厘米
术：喷塑、镂空、涂层
势：2014/2015秋冬海派流行趋势
器：服装面料
Name: Leopard Impression (3)
Conviction: Chinese blessing culture
Principle: approximation, interactive gradient, metal texture
Method: 4 sides in continuation, longitudinal tying back 64.2 cm
Operation: spraying, hollowing, coating
Trend: 2014/2015 autumn and winter Shanghai Fashion trend
Application : clothing fabrics

时间：2013年2月10日
地点：上海外滩
对象：外白渡桥
灵感：金属光泽、交错网格
Time: February 10, 2013
Location: the Bund, Shanghai
Object: the Waibaidu bridge
Inspiration: metallic luster, staggered grid

时间：2012年8月4日
地点：云南省博物馆
对象：出土金饰复制品
灵感：金属光泽、迷幻色彩
Time: August 4, 2012
Location: Museum of Yunnan
Object: unearthed gold jewelry replica
Inspiration: metallic luster, psychedelic

Creative Fabrics Design 1

名：视幻觉——山雨欲来风满楼（1）、（2）、（3）
道：道法自然、科技主义
理：近似重复、交互渐变、塑料质感
法：四方连缀、纬向接回21.6厘米
术：梭织提花、轧花、拔染、涂层
势：2013/2014国际流行趋势之地理形态
器：服装面料

Name: Visual Hallucinations – the Rising Wind Forebodes the Coming Storm (1), (2), (3)
Conviction: nature, science and technology
Principle: approximated duplicate, interactive gradient, plasticity
Method: 4 sides in continuation, zonal tying-in 21.6 cm
Operation: jacquard weaving, embossing, discharging, coating
Trend: 2013/2014 international fashion trend of geographic features
Application: clothing fabrics

Creative Fabrics Design 1

名：节节生（2）
道：民俗
理：偶发、抹灰质感、土墙质感
法：四方连缀、纬向接回35厘米
术：撕扯、腐蚀、双层粘合布
势：2013/2014国际流行民族文化形态
器：装饰布

Name: Transformation (2)
Conviction: folk
Principle: accidental, plastering texture, wall texture
Method: 4 sides in continuation, zonal tying-in 35 cm
Operation: tearing, corroding, double fabric bonding
Trend: 2013/2014 international fashion trend of folk culture
Application: decorative cloth

时间：2009年7月25日
地点：西藏拉萨
对象：布达拉宫白墙
灵感：经年累月、饱经沧桑
Time: July 25, 2009
Location: Lasa, Tibet
Object: white wall of Potala Palace
Inspiration: rich in time, rich in life

Creative Fabrics Design 1

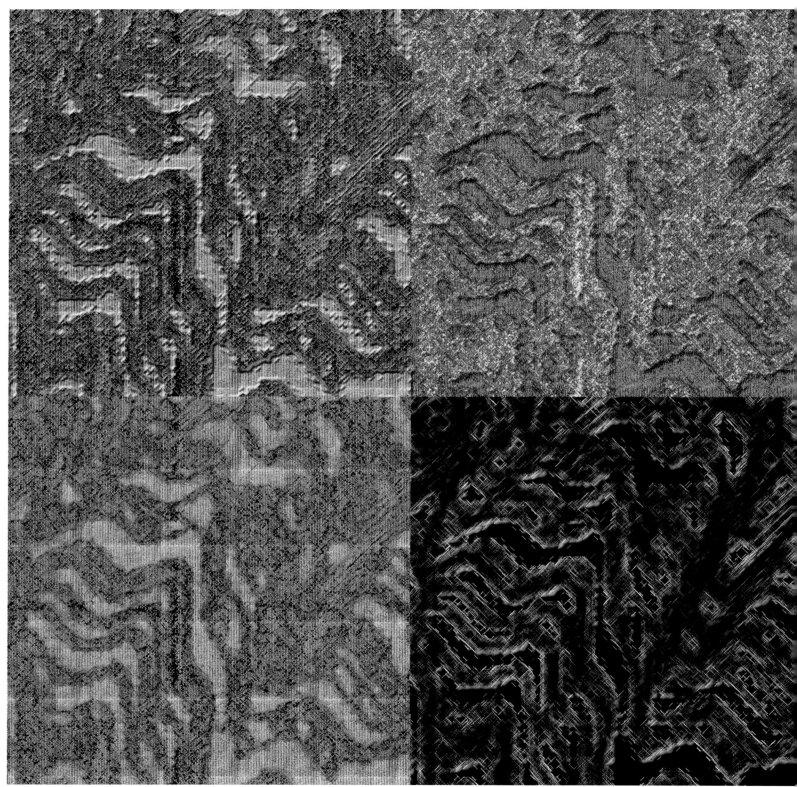

名：虎豹纹与植物纹混搭（3）、（5）、（6）、（7）
道：道法自然
理：近似、交互渐变、混搭
法：四方连缀、纬向接回35厘米
术：梭织提花、轧花、水洗
势：2013/2014国际流行趋势之生物形态
器：服装面料

Name: Mixture of Leopard Pattern and Plant Pattern (3), (5), (6), (7)
Conviction: imitation of nature
Principle: approximation, interactive gradient, mix and-match
Method: 4 sides in continuation, zonal tying-in 35 cm
Operation: jacquard weaving, embossing, washing
Trend: 2013/2014 international fashion trend of bio feature
Application: clothing fabrics

时间：2010年8月3日
地点：四川甘孜州
对象：水中石
灵感：湍急的河流中石头湿漉漉的质感
Time: August 3, 2010
Location: Ganzi, Sichuan
Object: a stone in the water
Inspiration: sense of moistness of stone in a rapid stream

Creative Fabrics Design 1

名：虎豹纹与植物纹混搭（8）
道：道法自然
理：近似、交互渐变、混搭
法：四方连缀、纬向接回21.6厘米
术：梭织提花、烧灼镂空
势：2013/2014国际流行趋势之生物形态
器：女装面料
Name: Mixture of Leopard Pattern and Plant Pattern(8)
Conviction: imitation of nature
Principle: approximation, interactive gradient, mix-and-match
Method: 4 sides in continuation, zonal tying-in 21.6 cm
Operation: jacquard weaving, burning and hollowing
Trend: 2013/2014 international fashion trend of bio feature
Application: women's fabric

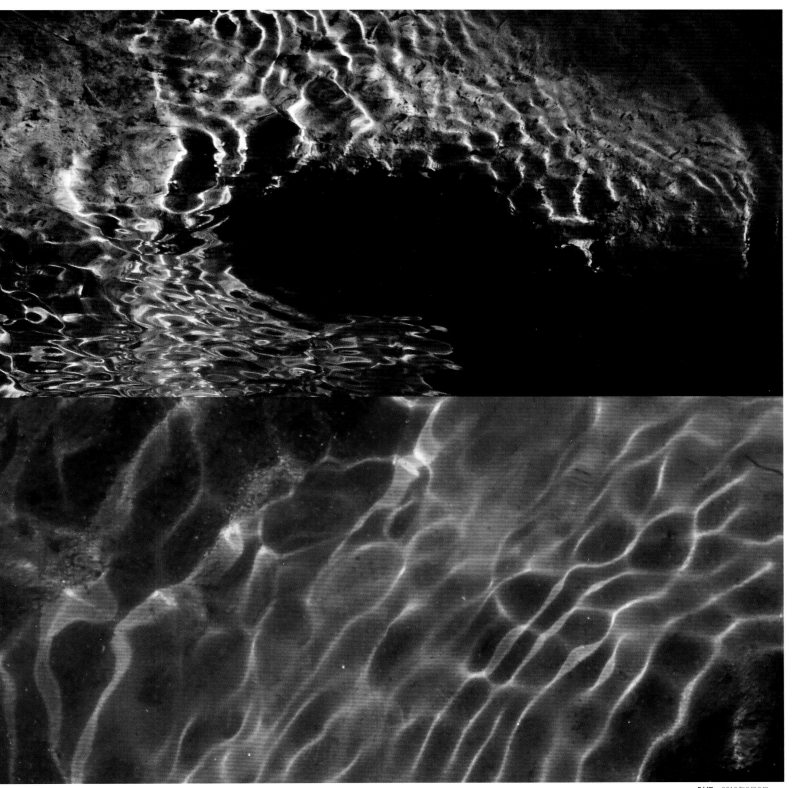

时间：2012年8月2日
地点：四川宜宾
对象：水纹
灵感：水中光斑
Time: August 2, 2012
Location: Yibin, Sichuan
Object: water
Inspiration: water wave

Creative Fabrics Design 1

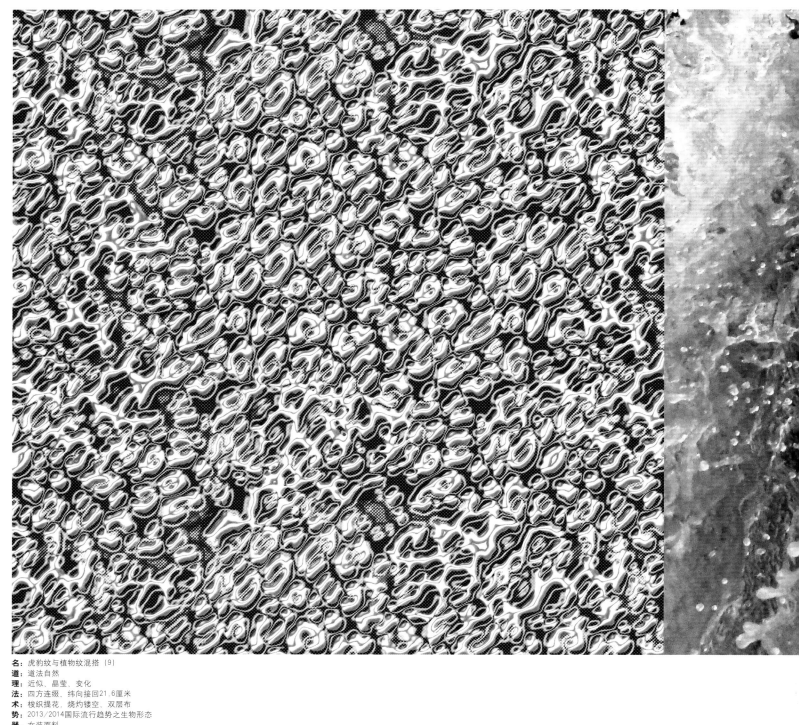

名：虎豹纹与植物纹混搭（9）
道：道法自然
理：近似，晶莹，变化
法：四方连缀、纬向接回21.6厘米
术：梭织提花，烧灼镂空，双层布
势：2013/2014国际流行趋势之生物形态
器：女装面料
Name: Mixture of Leopard Pattern and Plant Pattern (9)
Conviction: imitation of nature
Principle: approximation, glittering and translucent, change
Method: 4 sides in continuation, zonal tying-in 21.6 cm
Operation: jacquard weaving, burning, hollowing, cloth doubling
Trend: 2013/2014 international fashion trend of bio feature
Application: women's fabric

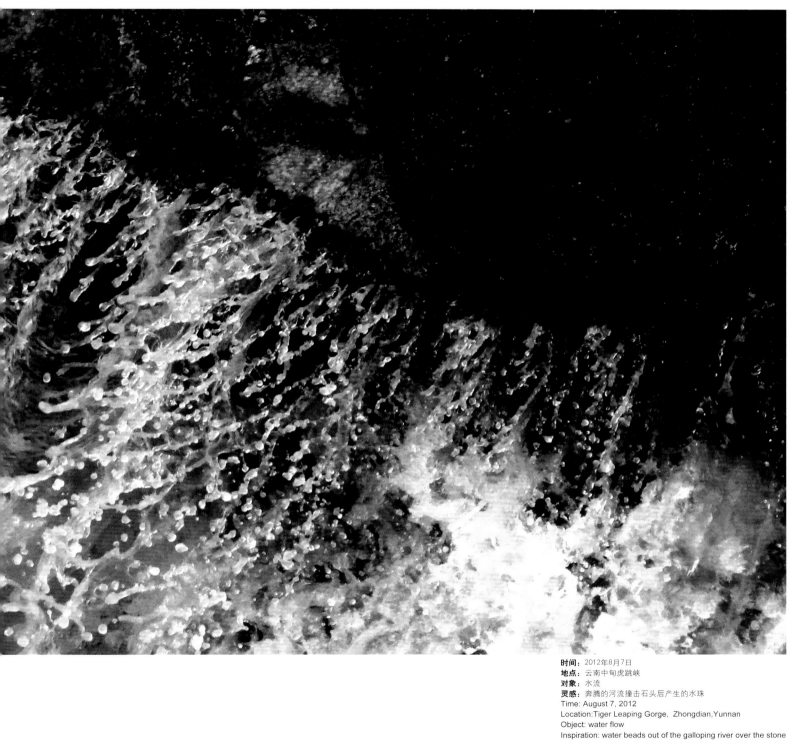

时间：2012年8月7日
地点：云南中甸虎跳峡
对象：水流
灵感：奔腾的河流撞击石头后产生的水珠
Time: August 7, 2012
Location: Tiger Leaping Gorge, Zhongdian, Yunnan
Object: water flow
Inspiration: water beads out of the galloping river over the stone

Creative Fabrics Design 1

名：虎豹纹与植物纹混搭（20）
道：道法自然
理：近似、晶莹、变化
法：四方连缀、纬向接回21.6厘米
术：梭织提花、烧灼镂空、双层布
势：2013/2014国际流行趋势之生物形态
器：女装面料

Name: Mixture of Leopard Pattern and Plant Pattern (20)
Conviction: imitation of nature
Principle: approximation, glittering and translucent, change
Method: 4 sides in continuation, zonal tying-in 21.6 cm
Operation: jacquard weaving, burning and hollowing, cloth doubling
Trend: 2013/2014 international fashion trend of bio feature
Application: women's fabric

时间：2012年11月7日
地点：浙江千岛湖
对象：蛇
灵感：交织缠绕的蛇、冷艳
Time: November 7, 2012
Location: Qiandao Lake, Zhejiang
Object: the snake
Inspiration: entwined snake, quiet and magnificent

Creative Fabrics Design 1

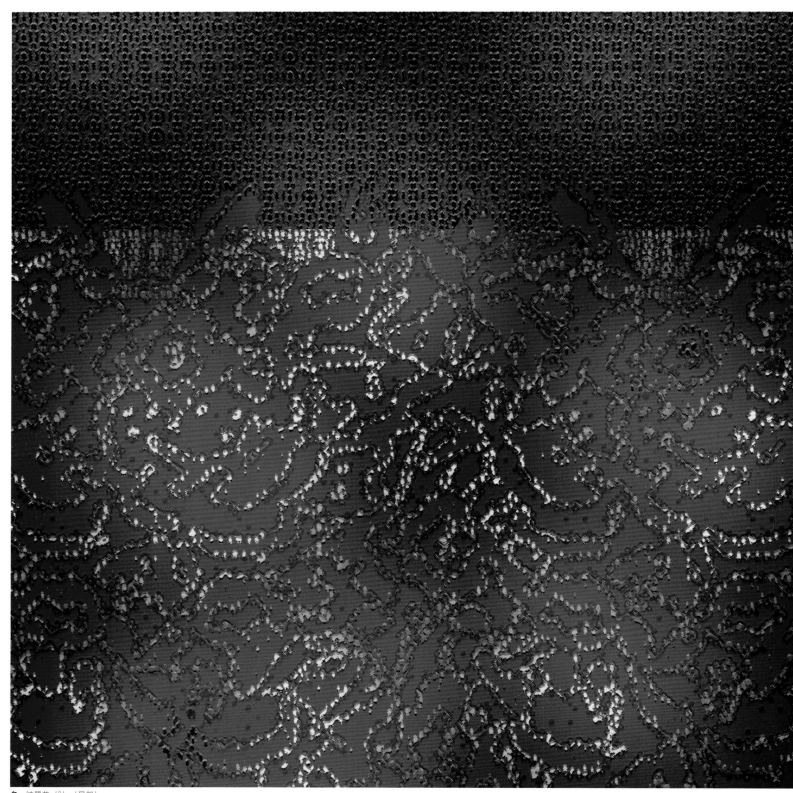

名：结盟花（3）（局部）
道：民俗
理：近似、交互无序到有序
法：四方连缀，纬向接回35厘米
术：镂空、轧花
势：2013/2014国际流行趋势之民族文化形态
器：女装面料

Name: The Allied Flower (3) (portion)
Conviction: folk
Principle: approximation, interaction from disorderly to orderly
Method: 4 sides in continuation, zonal tying-in 35 cm
Operation: hollowing, embossing
Trend: 2013/2014 international fashion trend of national culture
Application : women's fabric

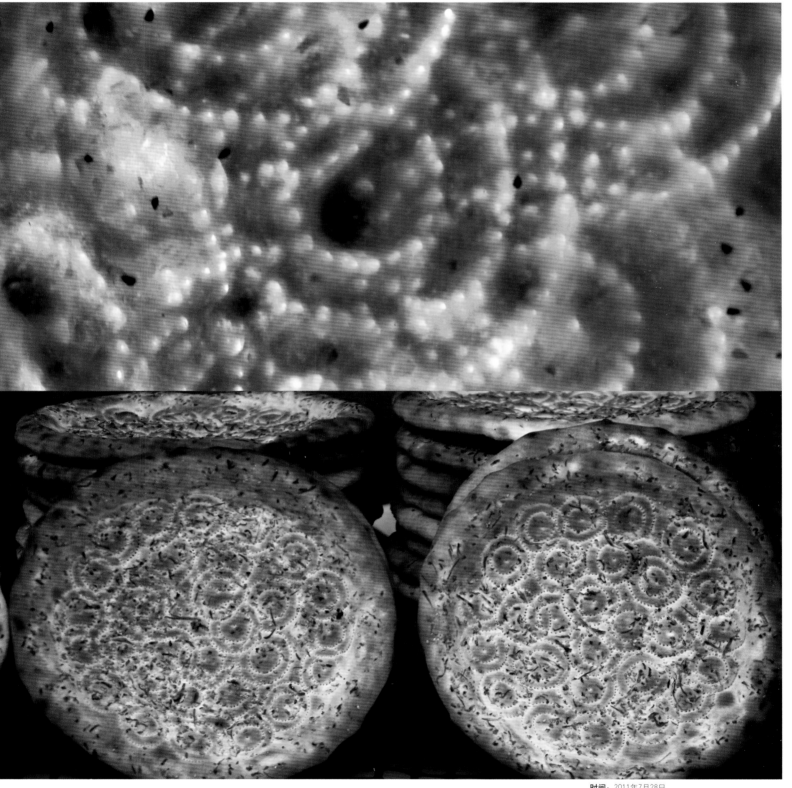

时间：2011年7月28日
地点：新疆和田大巴扎
对象：馕
灵感：古老西域，丝绸之路
Time: July 28, 2011
Location: bazaar, Hetian, Xinjiang
Object: nang — a kind of crusty pancake
Inspiration: the western regions, the ancient Silk Road

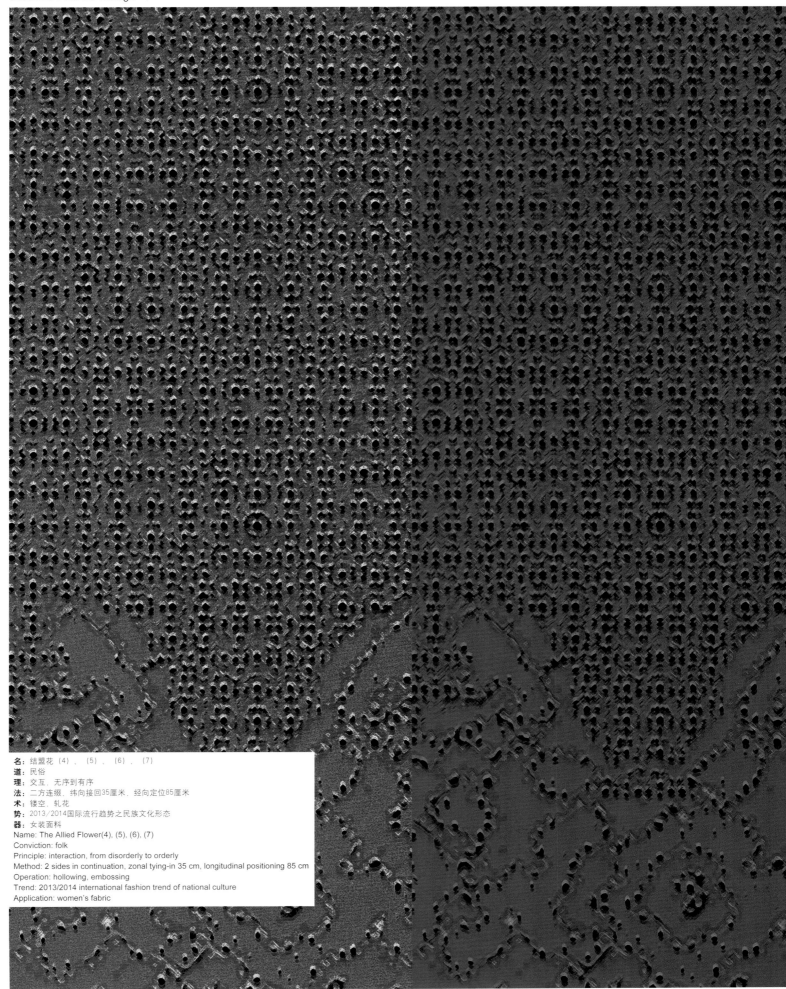

名：结盟花（4）、（5）、（6）、（7）
道：民俗
理：交互．无序到有序
法：二方连缀、纬向接回35厘米，经向定位85厘米
术：镂空．轧花
势：2013/2014国际流行趋势之民族文化形态
器：女装面料
Name: The Allied Flower(4), (5), (6), (7)
Conviction: folk
Principle: interaction, from disorderly to orderly
Method: 2 sides in continuation, zonal tying-in 35 cm, longitudinal positioning 85 cm
Operation: hollowing, embossing
Trend: 2013/2014 international fashion trend of national culture
Application: women's fabric

Creative Fabrics Design 1

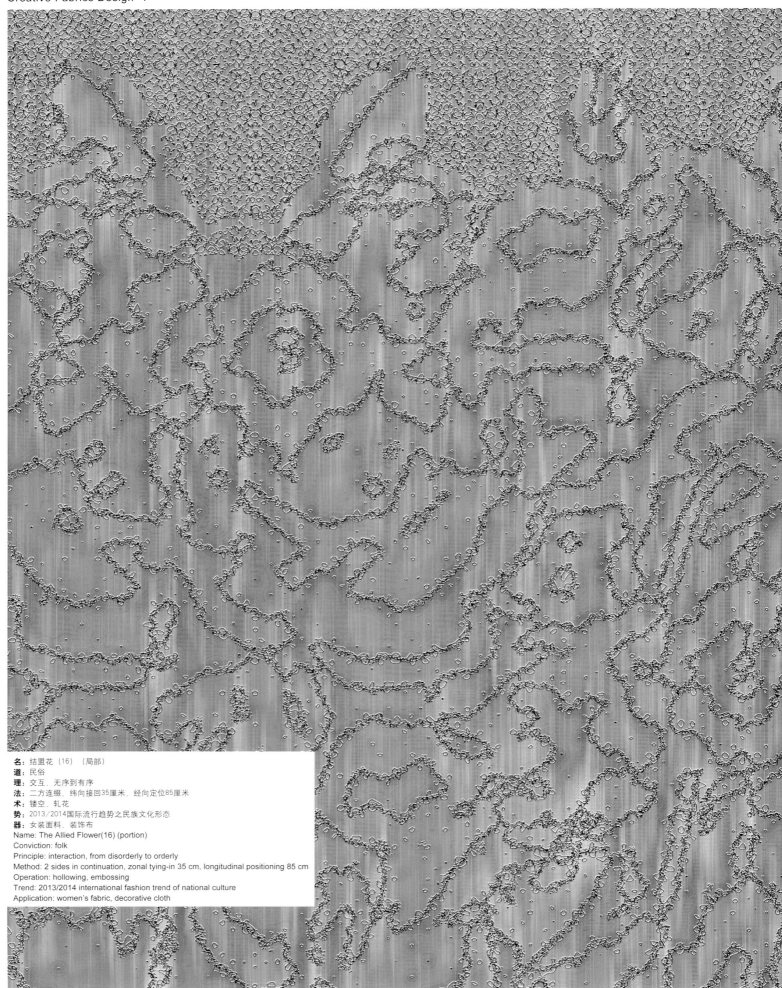

名：结盟花（16）（局部）
道：民俗
理：交互，无序到有序
法：二方连缀，纬向接回35厘米，经向定位85厘米
术：镂空、轧花
势：2013/2014国际流行趋势之民族文化形态
器：女装面料、装饰布

Name: The Allied Flower(16) (portion)
Conviction: folk
Principle: interaction, from disorderly to orderly
Method: 2 sides in continuation, zonal tying-in 35 cm, longitudinal positioning 85 cm
Operation: hollowing, embossing
Trend: 2013/2014 international fashion trend of national culture
Application: women's fabric, decorative cloth

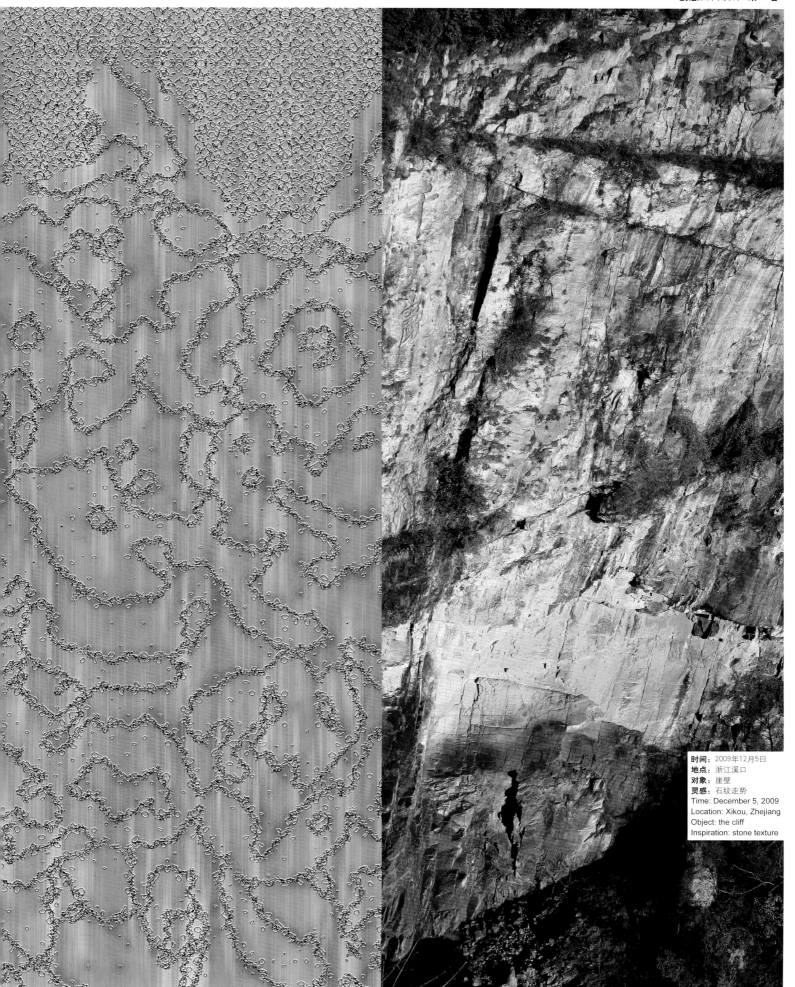

时间：2009年12月5日
地点：浙江溪口
对象：崖壁
灵感：石纹走势
Time: December 5, 2009
Location: Xikou, Zhejiang
Object: the cliff
Inspiration: stone texture

Creative Fabrics Design 1

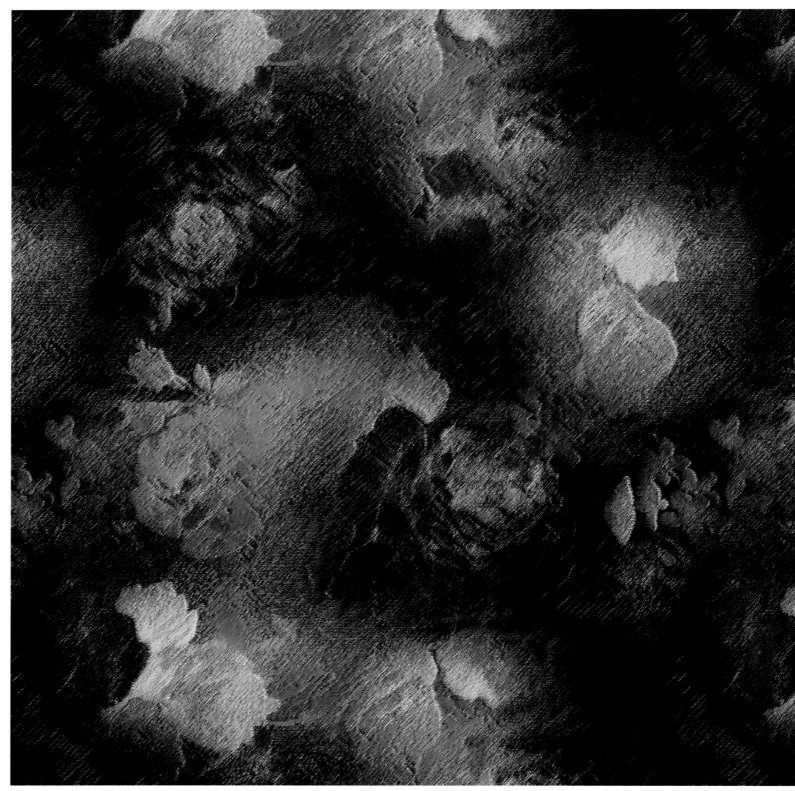

名：牡丹功夫（7）
道：美国好莱坞20世纪三四十年代电影
理：斑驳、德国蔡司镜头中的油润光影
法：纬向接回35厘米，经向定位85厘米
术：烂花、轧花
势：2013/2014国际流行趋势之怀旧形态
器：女装面料、装饰布

Name: Kung Fu Peony (7)
Conviction: Hollywood movies in1930 and 1940
Principle: mottled, the light and shade in German Zeiss lenses
Method: zonal tying-in 35 cm, longitudinal positioning 85 cm
Opration: a burnt-discharging, embossing
Trend: 2013/2014 international fashion trend of nostalgia
Application: women's fabric, decorative cloth

时间：2012年8月10日
地点：四川西昌
对象：光影下的石阶和红墙
灵感：影调
Time: August 10, 2012
Location: Xichang, Sichuan
Object: steps and walls in light and shade
Inspiration: shades of light

Creative Fabrics Design 1

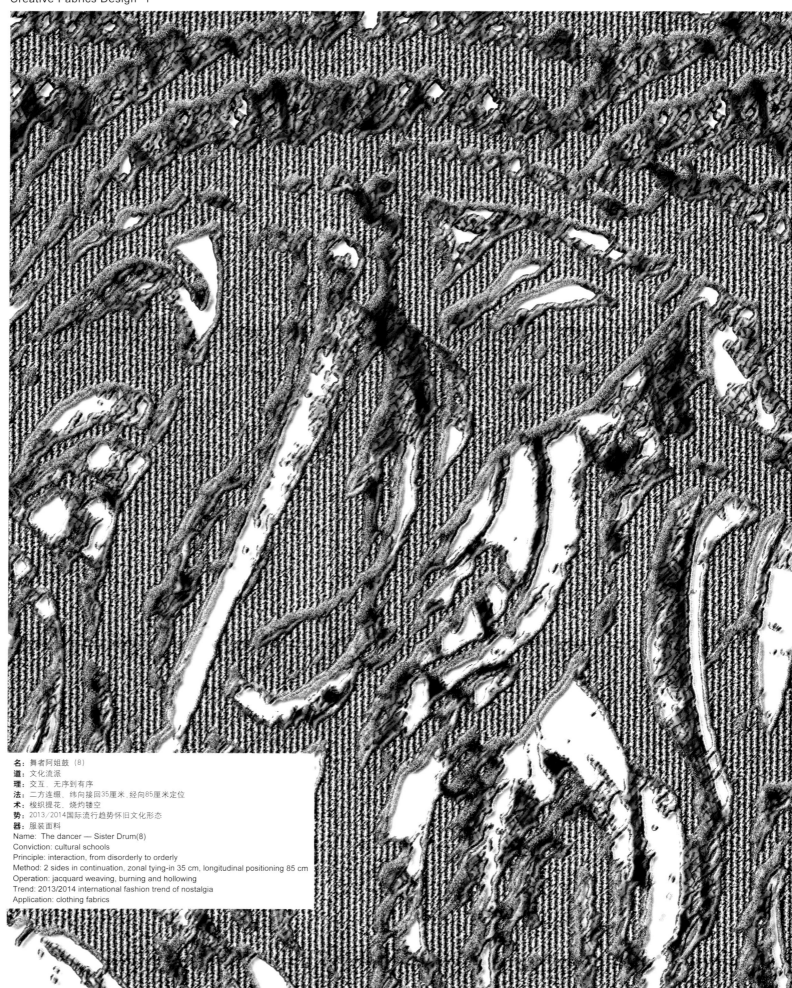

名：舞者阿姐鼓（8）
道：文化流派
理：交互，无序到有序
法：二方连缀，纬向接回35厘米，经向85厘米定位
术：梭织提花，烧灼镂空
势：2013/2014国际流行趋势怀旧文化形态
器：服装面料

Name: The dancer — Sister Drum(8)
Conviction: cultural schools
Principle: interaction, from disorderly to orderly
Method: 2 sides in continuation, zonal tying-in 35 cm, longitudinal positioning 85 cm
Operation: jacquard weaving, burning and hollowing
Trend: 2013/2014 international fashion trend of nostalgia
Application: clothing fabrics

时间：2011年2月14日
地点：坦桑尼亚桑给巴尔岛
对象：煤油灯
灵感：早期工业文明器具
Time: February 14, 2011
Location: Tanzania island of Zanzibar
Object: kerosene lamp
Inspiration: the apparatus in early industrial civilization

Creative Fabrics Design 1

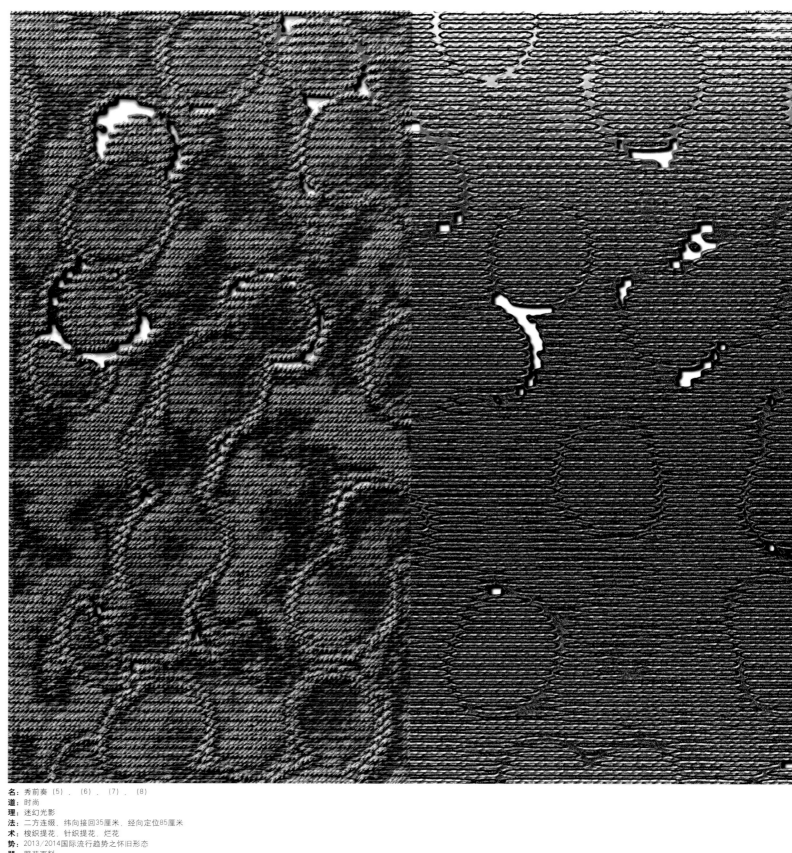

名：秀前奏（5）、（6）、（7）、（8）
道：时尚
理：迷幻光影
法：二方连缀、纬向接回35厘米，经向定位85厘米
术：梭织提花，针织提花，烂花
势：2013/2014国际流行趋势之怀旧形态
器：服装面料

Name: A Prelude to the Show (5), (6), (7), (8)
Conviction: fashion
Principle: psychedelic light
Method: 2 sides in continuation, zonal tying-in 35 cm, longitudinal positioning 85 cm
Operation: jacquard weaving, knitted fabric jacquard treating, burnt-discharging
Trend: 2013/2014 international fashion trend of nostalgia
Application: clothing fabrics

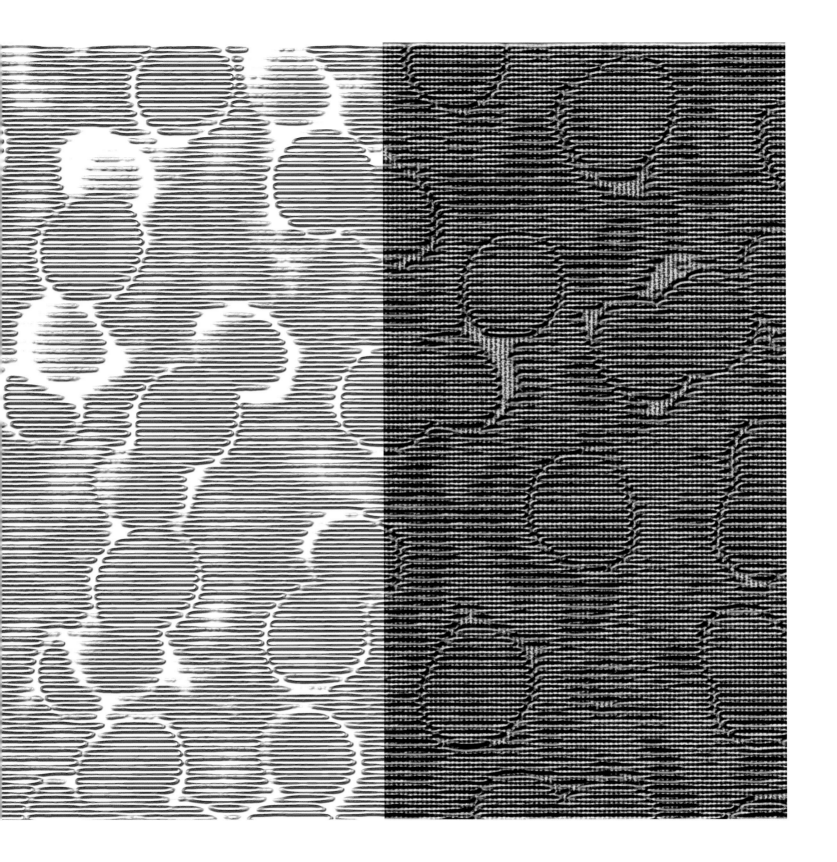

Creative Fabrics Design 1

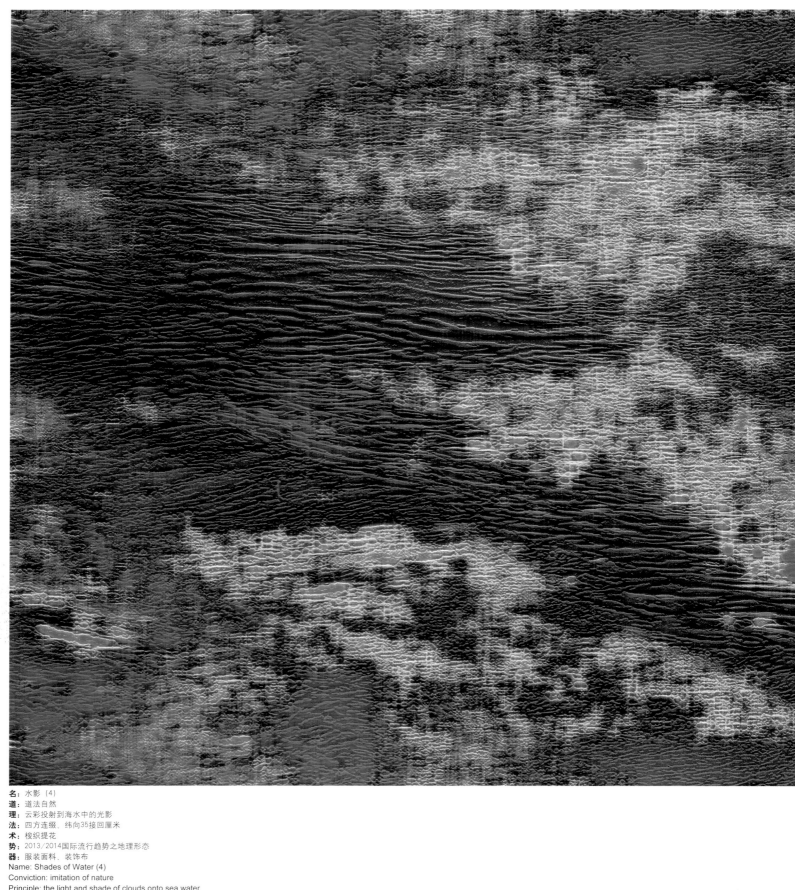

名：水影（4）
道：道法自然
理：云彩投射到海水中的光影
法：四方连缀、纬向35接回厘米
术：梭织提花
势：2013/2014国际流行趋势之地理形态
器：服装面料、装饰布

Name: Shades of Water (4)
Conviction: imitation of nature
Principle: the light and shade of clouds onto sea water
Method: 4 sides in continuation, zonal tying-in 35 cm
Operation: jacquard weaving
Trend: 2013/2014 international fashion trend of geo feature
Application: clothing fabrics, decorative cloth

时间：2011年2月18日
地点：坦桑尼亚桑给巴尔岛
对象：光影下的海水
灵感：变化中的影调和肌理
Time: February 18, 2011
Location: Island of Zanzibar, Tanzania
Object: sea water under arrays of light and shade
Inspiration: variations on the shades and texture

Creative Fabrics Design 1

名：灵璧石 (4)、(5)
道：道法自然
理：石头在激流中经过打磨后的润泽
法：二方连缀，纬向接回29厘米，经向定位85厘米
术：梭织提花
势：2013/2014国际流行趋势之地理形态
器：服装面料

Name: Lingbi Stone (4), (5)
Conviction: imitation of nature
Principle: mellow texture of stone polishing in the rapids
Method: 2 sides in continuation, zonal tying-in 29 cm, longitudinal positioning 85 cm
Operation: jacquard weaving
Trend: 2013/2014 international fashion trend of geo feature
Application: clothing fabrics

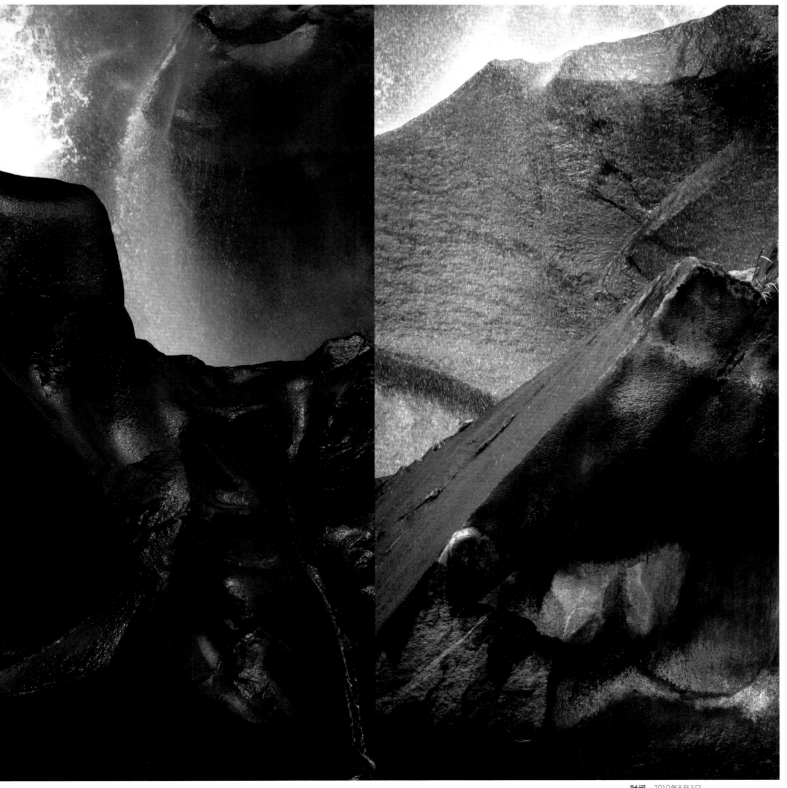

时间：2010年8月3日
地点：四川川藏公路旁
对象：激流下的石头
灵感：灵璧石
Time: August 3, 2010;
Location: Sichuan-Tibet Highway, Sichuan
Object: stone in rapids
Inspiration: Lingbi stone

Creative Fabrics Design 1

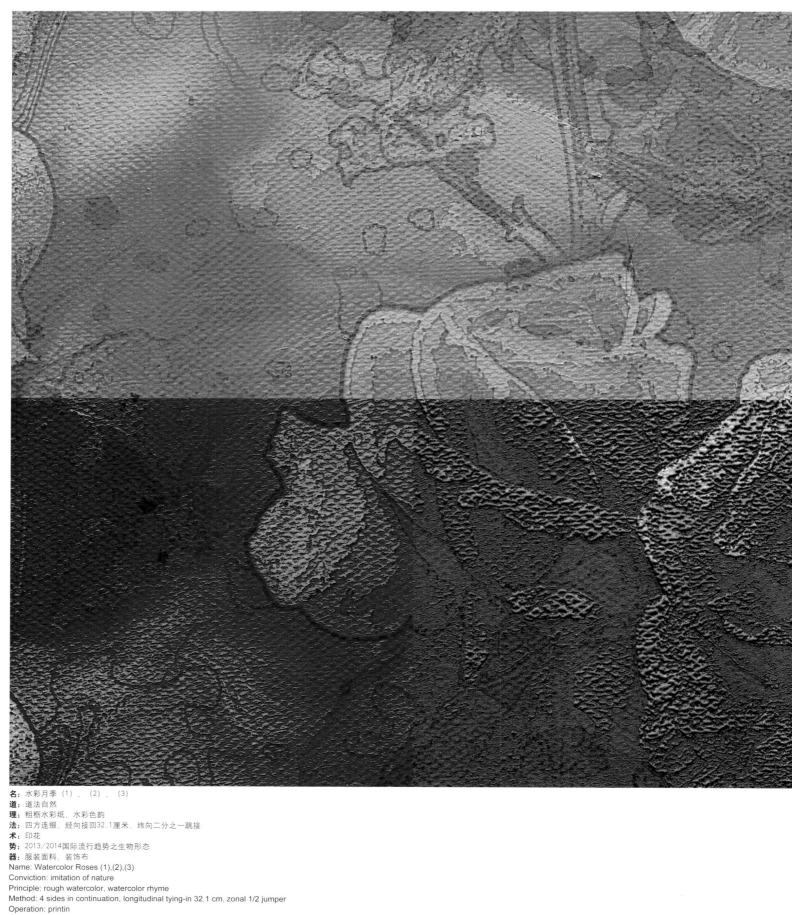

名：水彩月季（1）、（2）、（3）
道：道法自然
理：粗粝水彩纸、水彩色韵
法：四方连缀，经向接回32.1厘米，纬向二分之一跳接
术：印花
势：2013/2014国际流行趋势之生物形态
器：服装面料、装饰布

Name: Watercolor Roses (1),(2),(3)
Conviction: imitation of nature
Principle: rough watercolor, watercolor rhyme
Method: 4 sides in continuation, longitudinal tying-in 32.1 cm, zonal 1/2 jumper
Operation: printin
Trend: 2013/2014 international fashion trend of bio feature
Conviction: clothing fabrics, decorative cloth

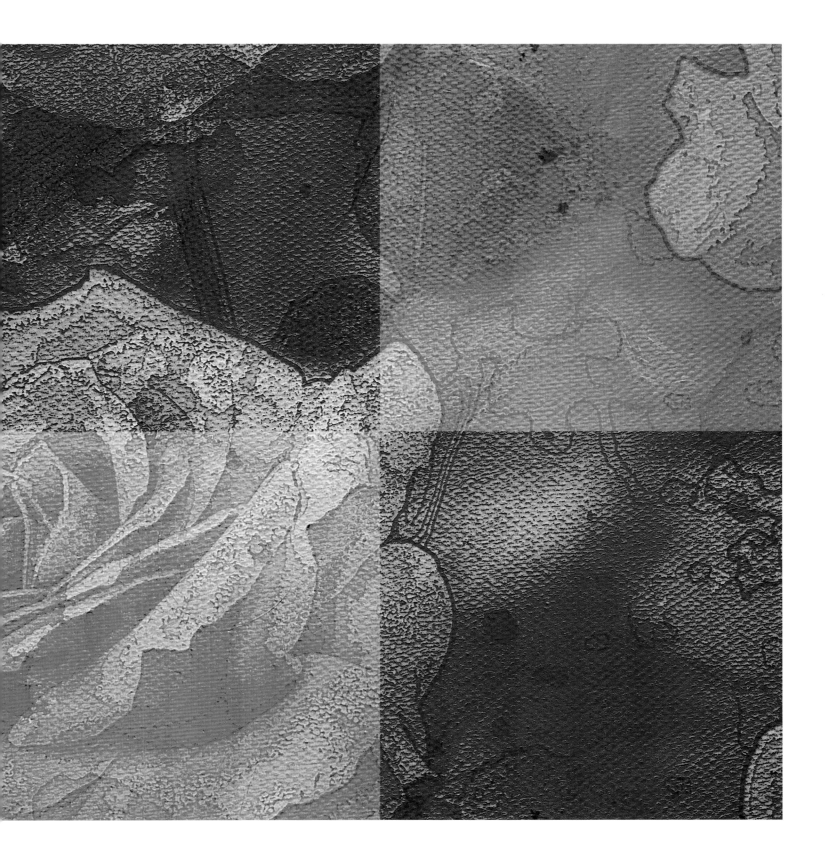

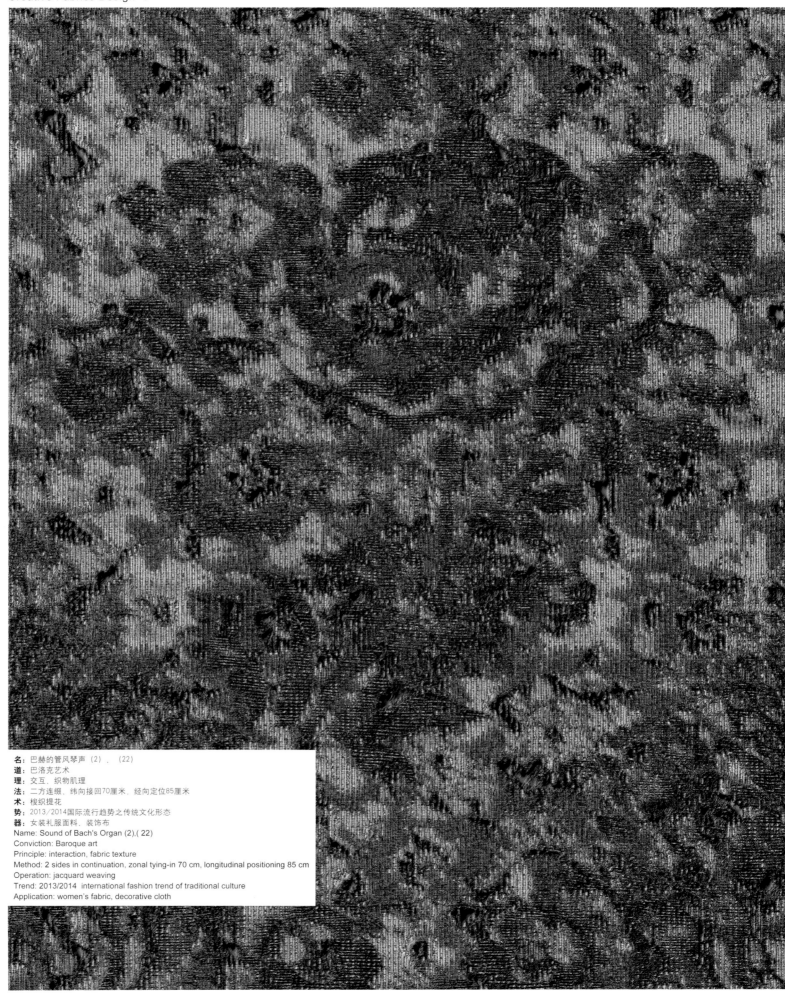

名：巴赫的管风琴声（2）、（22）
道：巴洛克艺术
理：交互、织物肌理
法：二方连缀，纬向接回70厘米，经向定位85厘米
术：梭织提花
势：2013/2014国际流行趋势之传统文化形态
器：女装礼服面料、装饰布

Name: Sound of Bach's Organ (2),(22)
Conviction: Baroque art
Principle: interaction, fabric texture
Method: 2 sides in continuation, zonal tying-in 70 cm, longitudinal positioning 85 cm
Operation: jacquard weaving
Trend: 2013/2014 international fashion trend of traditional culture
Application: women's fabric, decorative cloth

Creative Fabrics Design 1

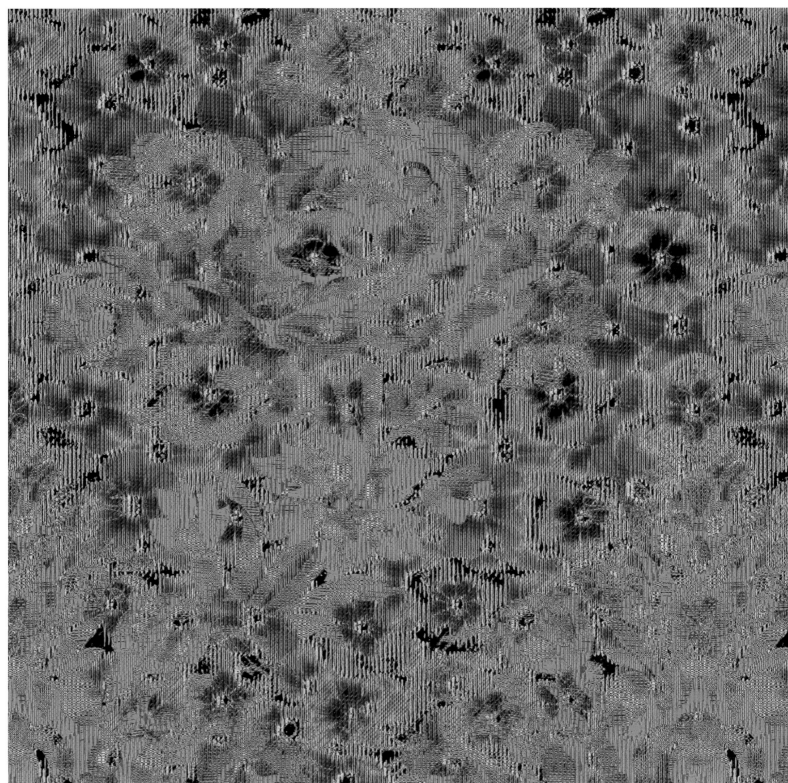

名：巴赫的管风琴声（6）、（9）
道：巴洛克艺术
理：织物及刺绣肌理、晕染
法：二方连缀，纬向接回70厘米，经向定位85厘米
术：梭织提花，刺绣
势：2013/2014国际流行趋势之传统文化形态
器：女装礼服面料、装饰布

Name: Sound of Bach's Organ (6), (9)
Conviction: Baroque art
Principle: fabrics and embroidery fabric, halo
Method: 2 sides in continuation, zonal tying-in 70 cm, longitudinal positioning 85cm
Operation: jacquard weaving, embroidery
Trend: 2013/2014 international fashion trend of traditional culture
Application: women's fabric, decorative cloth

Creative Fabrics Design 1

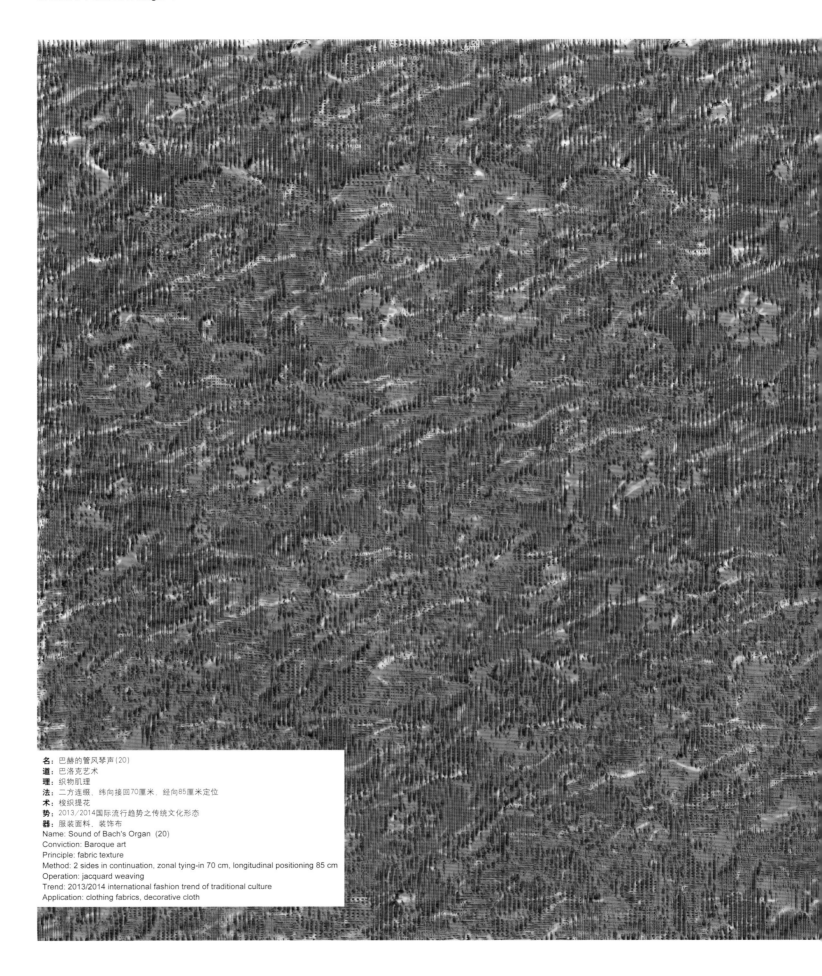

名：巴赫的管风琴声(20)
道：巴洛克艺术
理：织物肌理
法：二方连缀、纬向接回70厘米、经向85厘米定位
术：梭织提花
势：2013/2014国际流行趋势之传统文化形态
器：服装面料、装饰布

Name: Sound of Bach's Organ (20)
Conviction: Baroque art
Principle: fabric texture
Method: 2 sides in continuation, zonal tying-in 70 cm, longitudinal positioning 85 cm
Operation: jacquard weaving
Trend: 2013/2014 international fashion trend of traditional culture
Application: clothing fabrics, decorative cloth

时间：2010年10月20日
地点：尼泊尔博卡拉
对象：石墙
灵感：手工堆砌的不经意感
Time: October 20, 2010
Location: Pokhara, Nepal
Object: stone wall
Inspiration: inadvertent feeling of manual stacked wall

Creative Fabrics Design 1

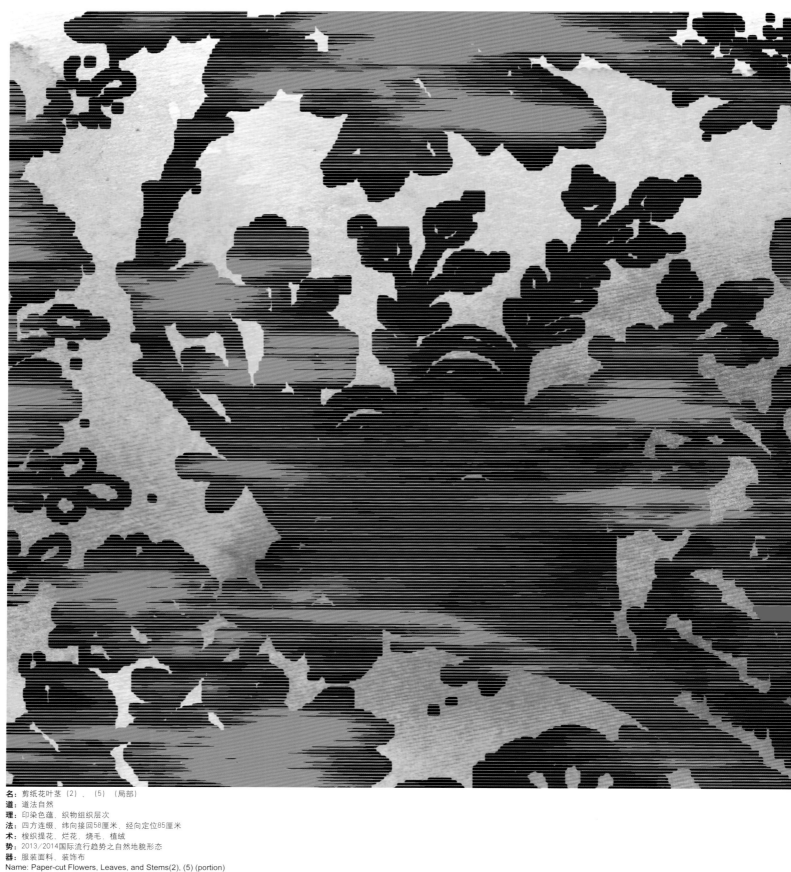

名：剪纸花叶茎（2）、（5）（局部）
道：道法自然
理：印染色蕴，织物组织层次
法：四方连缀，纬向接回58厘米，经向定位85厘米
术：梭织提花，烂花，烧毛，植绒
势：2013/2014国际流行趋势之自然地貌形态
器：服装面料，装饰布

Name: Paper-cut Flowers, Leaves, and Stems(2), (5) (portion)
Conviction: imitation of nature
Principle: printing and dyeing color, fabric texture
Method: 4 sides in continuation, zonal tying-in 58 cm, longitudinal postioning 85 cm
Operation: jacquard weaving, burnt-discharging, singeing, flocking
Trend: 2013/2014 international fashion trend of natural landform
Application: clothing fabrics, decorative cloth

Creative Fabrics Design 1

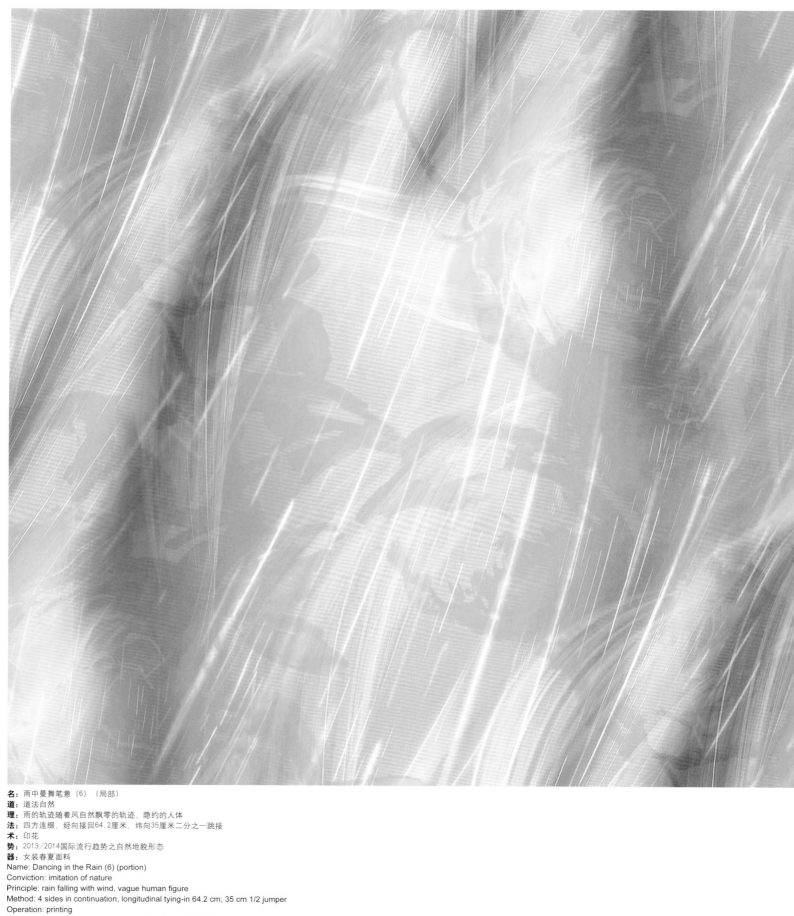

名：雨中曼舞笔意（6）（局部）
道：道法自然
理：雨的轨迹随着风自然飘零的轨迹、隐约的人体
法：四方连缀、经向接回64.2厘米、纬向35厘米二分之一跳接
术：印花
势：2013/2014国际流行趋势之自然地貌形态
器：女装春夏面料

Name: Dancing in the Rain (6) (portion)
Conviction: imitation of nature
Principle: rain falling with wind, vague human figure
Method: 4 sides in continuation, longitudinal tying-in 64.2 cm, 35 cm 1/2 jumper
Operation: printing
Trend: 2013/2014 international fashion trend of natural landform
Application: women's summer fabric

时间：2010年8月3日
地点：四川川藏公路旁
对象：水中石
灵感：石头在激流中的光泽
Time: August 3, 2010
Location: Sichuan-Tibet highway, Sichuan
Object: stone in the water
Inspiration: gloss of stone in rapids

Creative Fabrics Design 1

名：混血莎靓奥特（7）、（8）、（18）、（20）
道：道法自然
理：多种织物工艺交替使用及比较，斑驳颗粒或绒质
法：四方连缀，纬向接回35厘米
术：梭织提花、烂花、机绣、烧毛、植绒
势：2013/2014国际流行趋势之生物形态
器：服装面料、装饰布
Name: Half-blood Sha Liang Ott(7), (8), (18), (20)
Conviction: imitation of nature
Principle: alternating and comparing of a variety of fabrics, mottled particles or fleece
Method: 4 sides in continuation, zonal tying-in 35 cm
Operation: jacquard weaving, burnt-discharging, embroidery, singeing, flocking
Trend: 2013/2014 international fashion trend of bio feature
Application: clothing fabrics, decorative cloth

Creative Fabrics Design 1

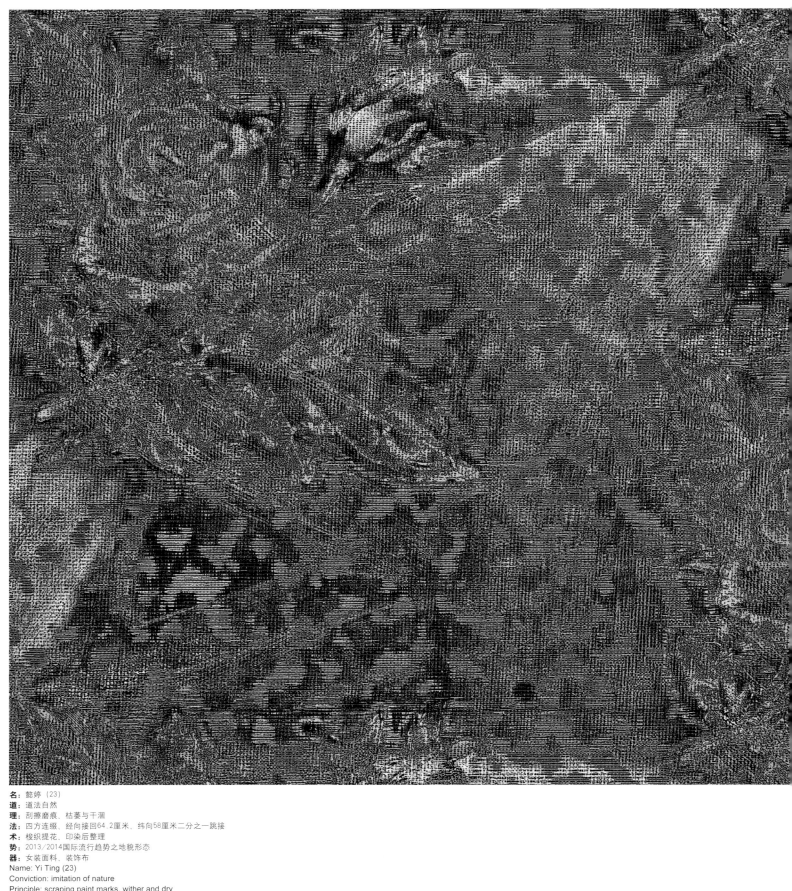

名：懿婷（23）
道：道法自然
理：刮擦磨痕、枯萎与干涸
法：四方连缀、经向接回64.2厘米、纬向58厘米二分之一跳接
术：梭织提花、印染后整理
势：2013/2014国际流行趋势之地貌形态
器：女装面料、装饰布
Name: Yi Ting (23)
Conviction: imitation of nature
Principle: scraping paint marks, wither and dry
Method: 4 sides in continuation, longitudinal tying-in 64.2 cm, 58 cm 1/2 jumper
Operation: jacquard weaving, printing and dyeing and finishing
Trend: 2013/2014 international fashion trend of geo-feature
Application: women's fabric, decorative cloth

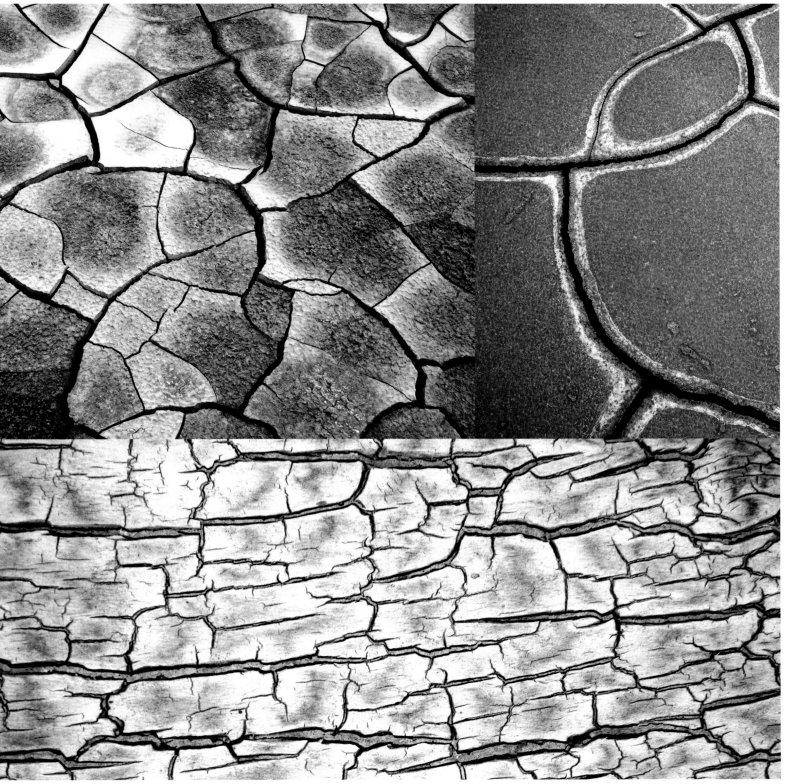

时间：2011年7月26日
地点：新疆塔克拉玛干沙漠
对象：盐碱地、木头
灵感：龟裂和干涸感
Time: July 26, 2011
Location: Taklimakan Desert, Xinjiang
Object: alkali soil, wood
Inspiration: cracking and dry

Creative Fabrics Design 1

名：建筑欧风之花雕（2）
道：文化流派
理：漏透，晕染
法：二方连缀，纬向接回35厘米，经向85厘米定位
术：梭织提花，腐蚀，染整
势：2013/2014国际流行趋势之传统文化形态
器：女装面料，装饰布
Name: European Style Architecture—Carving (2)
Conviction: cultural schools
Principle: hollowing, shading
Method: 2 sides in continuation, zonal tying-in 35 cm, longitudinal positioning 85 cm
Operation: jacquard weaving, corroding, dyeing and finishing
Trend: 2013/2014 international fashion trend of traditional culture
Application: women's fabric, decorative cloth

时间：2011年2月14日
地点：坦桑尼亚桑给巴尔岛
对象：木雕门
灵感：木刻浮雕
Time: February 14, 2011
Location: Island of Zanzibar, Tanzania
Objects: carved wood door
Inspiration: anaglyph

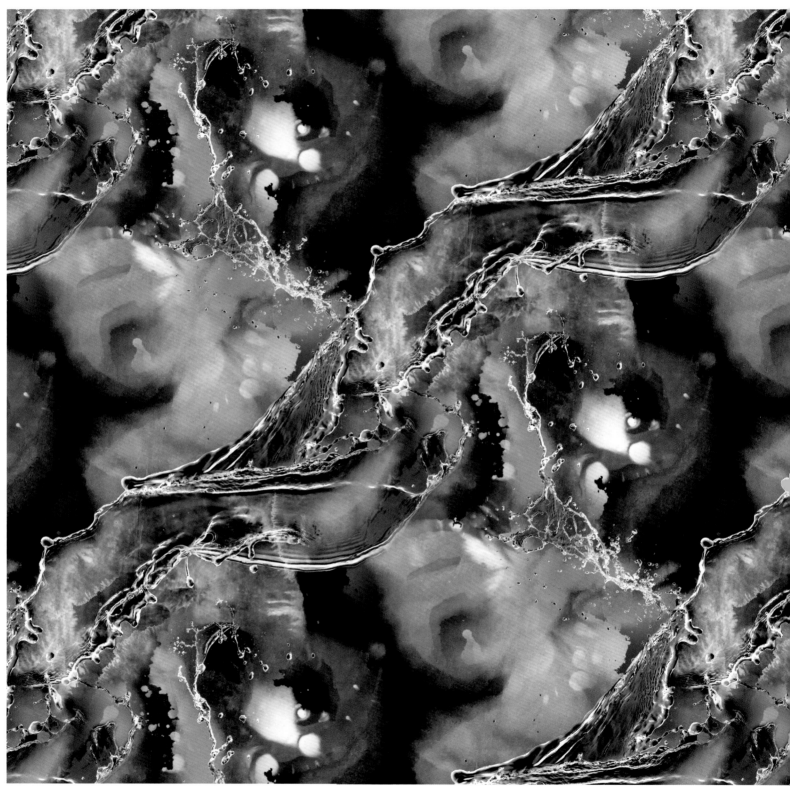

名：虎踞龙盘（2）
道：中国道家五行文化
理：晶莹剔透、水势奔腾
法：四方连缀，经向接回64.2厘米，纬向接回35厘米
术：转移印花
势：2013/2014国际流行趋势之传统文化形态
器：服装面料、装饰布

Name: Magnificent and Forbidding (2)
Conviction: the five elements of Chinese Taoist culture
Principle: crystal clear water, galloping water
Method: 4 sides in continuation, longitudinal back tying 64.2 cm, zonal tying-in 35 cm
Method: transfer printing
Trend: 2013/2014 international fashion trend of traditional culture
Application: clothing fabrics, decorative cloth

创意面料设计 第一卷

时间：2010年8月3日
地点：四川川藏公路旁
对象：河流
灵感：山涧、大河奔流、激流
Time: August 3, 2010
Location: Sichuan-Tibet Highway, Sichuan
Object: river
Inspiration: mountains, rivers, and rapids

Creative Fabrics Design 1

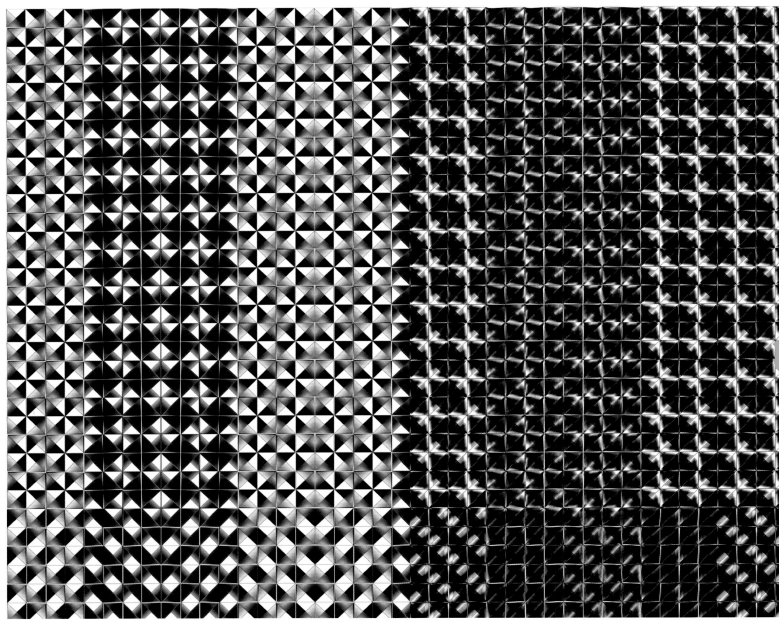

名：网格渐变坠落（1）、（2）、（3）、（4）
道：民俗文化
理：编织结构
法：二方连缀、纬向接回70厘米，经向定位88厘米
术：梭织提花、轧光、涂层
势：2014/2015秋冬海派流行趋势
器：服装面料；
Name: Grid in Gradation and Fall (1), (2), (3), (4)
Conviction: folk culture
Principle: knitting structure
Method: 2 sides in continuation, zonal tying-in 70 cm, longitudinal positioning 88 cm
Operation: jacquard weaving, calendering, coating
Trend: 2014/2015 autumn and winter fashion trend of Shanghai
Conviction: clothing fabrics

Creative Fabrics Design 1

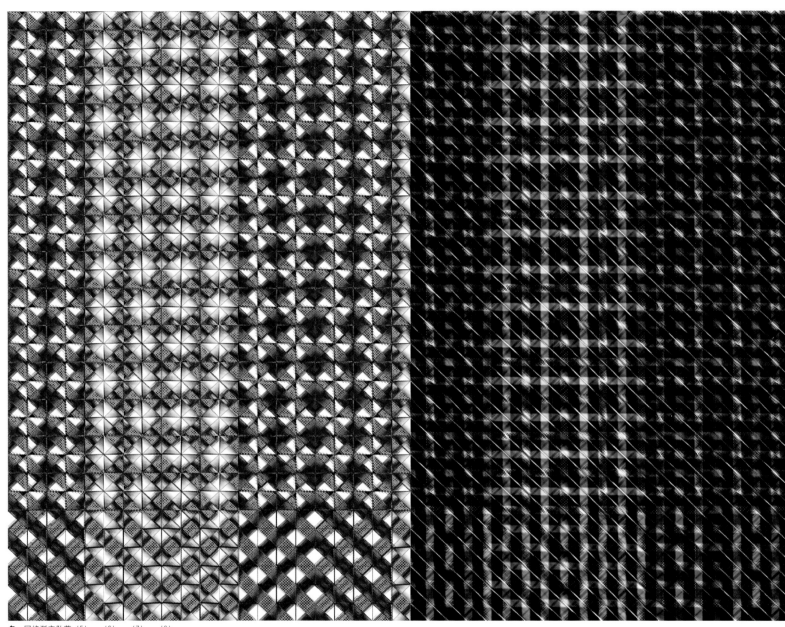

名：网格渐变坠落（5）、（6）、（7）、（8）
道：民俗文化
理：编织结构
法：二方连缀、纬向接回70厘米，经向定位88厘米
术：梭织提花、轧光、涂层
势：2014/2015秋冬海派流行趋势
器：服装面料
Name: Grid in Gradation and Fall (5), (6), (7), (8)
Conviction: folk culture
Principle: knitting structure
Method: 2 sides in continuation, zonal tying-in 70 cm, longitudinal positioning 88 cm
Operation: jacquard weaving, calendering, coating
Trend: 2014/2015 autumn and winter fashion trend of Shanghai
Application: clothing fabrics

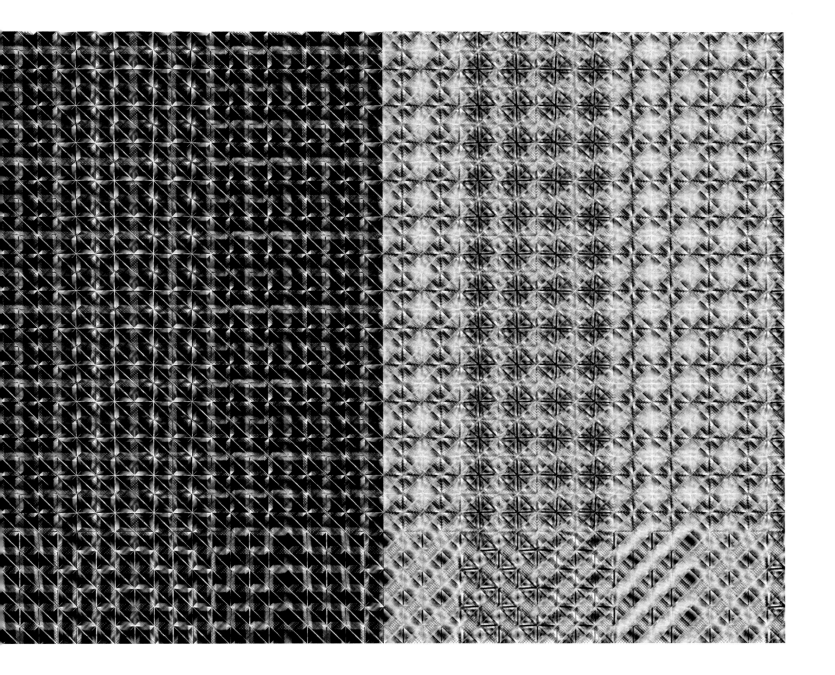

Creative Fabrics Design 1

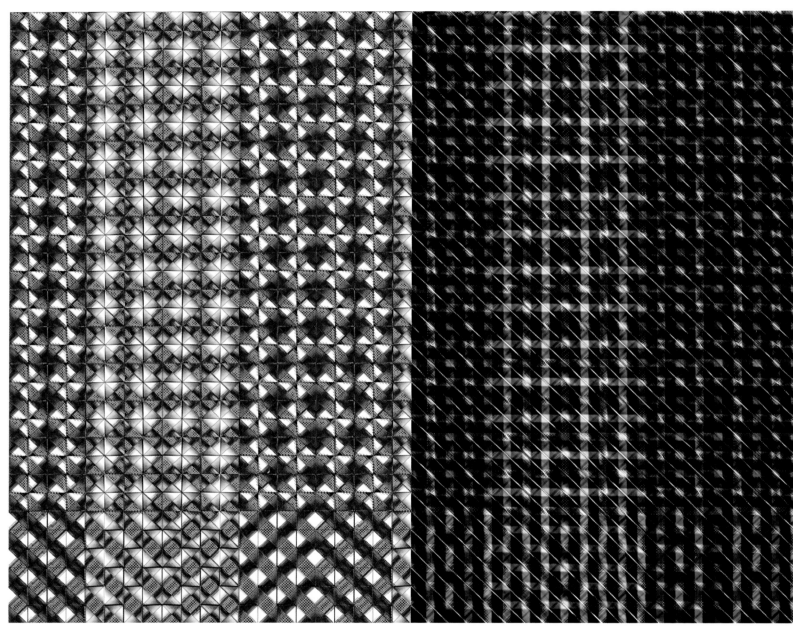

名：网格渐变坠落（9）、（10）、（11）、（12）
道：民俗文化
理：编织结构
法：二方连缀、纬向接回70厘米，经向定位88厘米
术：梭织提花、轧光、涂层
势：2014/2015秋冬海派流行趋势
器：服装面料

Name: Grid in Gradation and Fall (9), (10), (11), (12)
Conviction: folk culture
Principle: knitting structure
Method: 2 sides in continuation, zonal tying-in 70 cm, longitudinal positioning 88 cm
Operation: jacquard weaving, calendering, coating
Trend: 2014/2015 autumn and winter fashion trend of Shanghai
Application: clothing fabrics

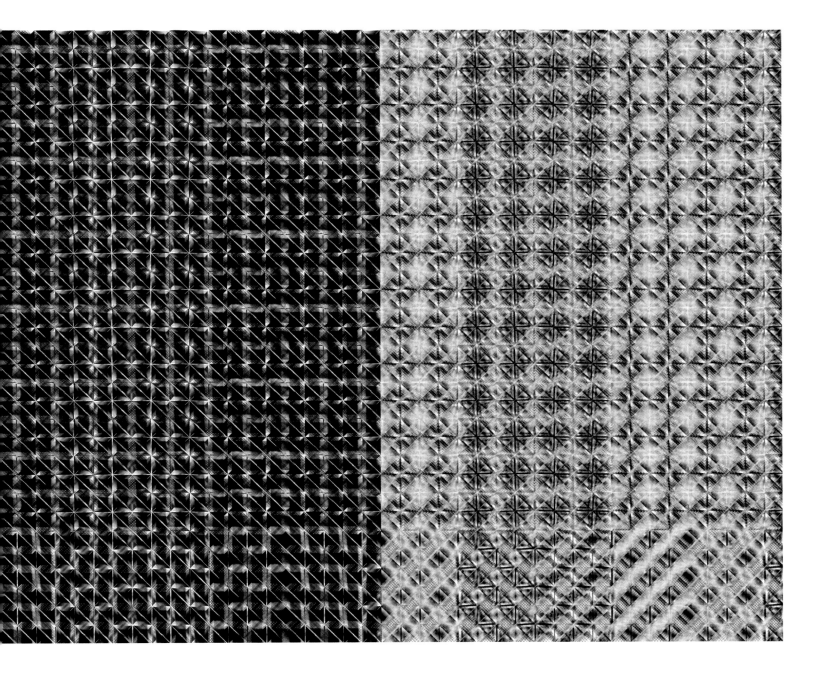

Creative Fabrics Design 1

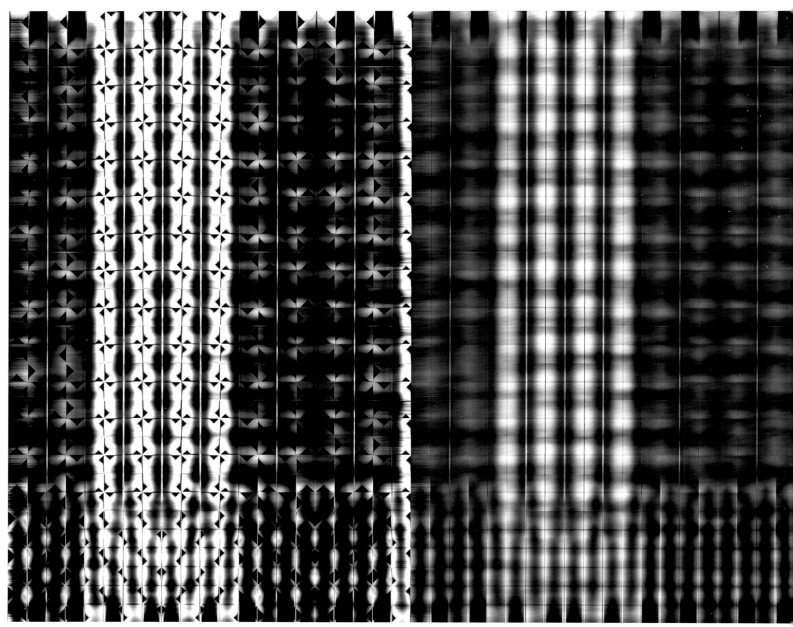

名： 网格渐变坠落（13）、（14）、（19）、（26）
道： 民俗文化
理： 编织结构
法： 二方连缀，纬向接回70厘米，经向定位88厘米
术： 梭织提花、轧光、涂层
势： 2014/2015秋冬海派流行趋势
器： 服装面料

Name: Grid in Gradation and Fall（13）,（14）,（19）,（26）
Conviction: folk culture
Principle: knitting structure
Method: 2 sides in continuation, zonal tying-in 70 cm, longitudinal positioning 88 cm
Operation: jacquard weaving, calendering, coating
Trend: 2014/2015 autumn and winter fashion trend of Shanghai
Application: clothing fabrics

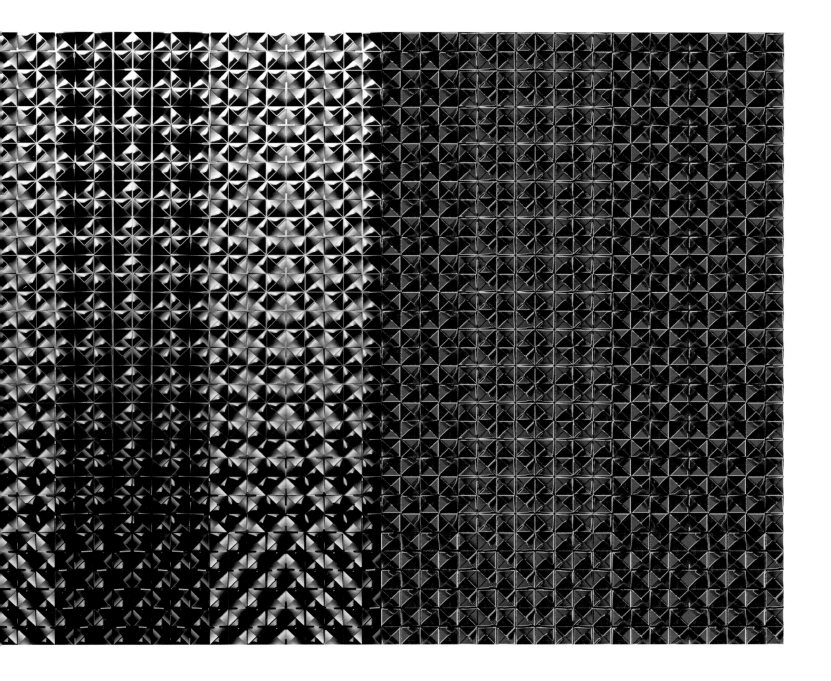

Creative Fabrics Design 1

名：网格渐变坠落（21）、（22）、（23）
道：民俗文化
理：编织结构
法：二方连缀、纬向接回70厘米，经向定位88厘米
术：梭织提花、轧光、涂层
势：2014/2015秋冬海派流行趋势
器：服装面料

Name: Grid in Gradation and Fall (21), (22), (23)
Conviction: folk culture
Principle: knitting structure
Method: 2 sides in continuation, zonal tying-in 70 cm, longitudinal positioning 88 cm
Operation: jacquard weaving, calendering, coating
Trend: 2014/2015 autumn and winter fashion trend of Shanghai
Application: clothing fabrics

时间：2010年11月29日
地点：上海淮海东路黄陂路口商厦
对象：金属链子
灵感：灯光下的光泽、熠熠生辉
Time: November 29, 2010
Location: shopping mall of Huaihai East Road and Huangpi Road, Shanghai
Object: metal chain
Inspiration: luster in the light, shining

Creative Fabrics Design 1

名：百乐门金里奇（1）（整体与局部）
道：爵士乐
理：水墨晕、韵
法：四方连缀，纬向接回58厘米，经向接回101.7厘米
术：印花，梭织提花
势：2013/2014国际流行趋势之艺术家表现形态
器：服装面料、装饰布
Name: Bailemen Gingrich (1) (whole and portion)
Conviction: jazz
Principle: ink shading, rhyme
Method: 4 sides in continuation, zonal tying-in 58 cm, longitudinal tying-in 101.7 cm
Operation: printing, jacquard weaving
Trend: 2013/2014 international fashion trend of artists expression
Application: clothing fabrics, decorative cloth

Creative Fabrics Design 1

名：上海老克拉（1）（整体与局部）
道：中国画审美观、爵士乐
理：水墨晕、韵
法：四方连缀，纬向接回58厘米，经向接回101.7厘米
术：印花，梭织提花
势：2013/2014国际流行趋势之艺术家表现形态
器：服装面料、装饰布
Name: Shanghai Old Carat (1) (whole and portion)
Conviction: Chinese painting aesthetics, jazz
Principle: ink shading, rhyme
Method: 4 sides in continuation, zonal tying-in 58 cm, longitudinal tying-in 101.7 cm
Operation: printing, jacquard weaving
Trend: 2013/2014 international fashion trend of artists expression
Device: clothing fabrics, decorative cloth

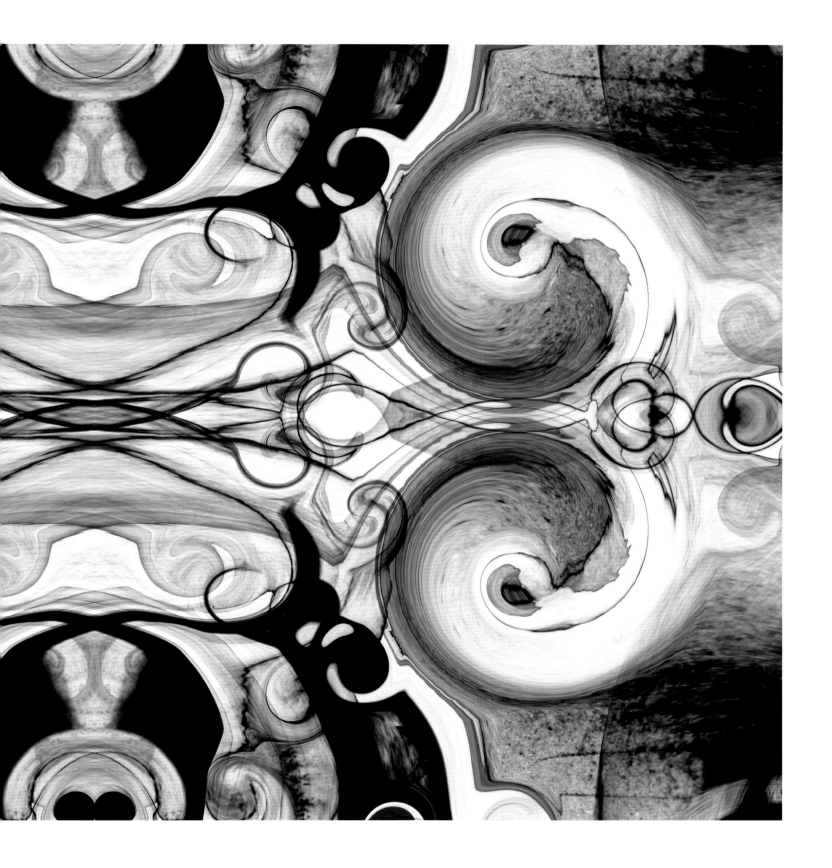

Creative Fabrics Design 1

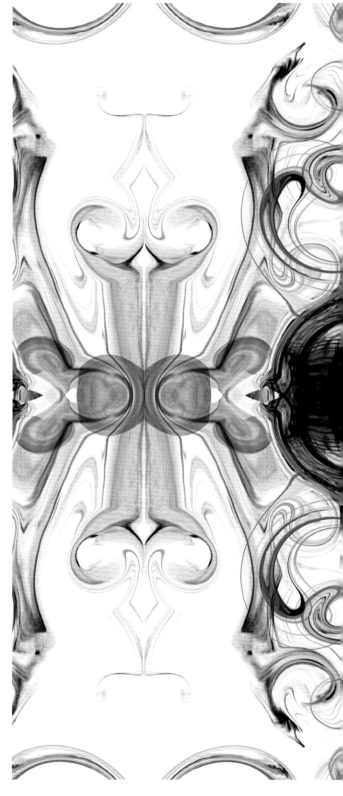

名：水墨宫庭花纹梦露之梦成（1）
道：中国画审美观，爵士乐
理：水墨晕、韵
法：四方连缀，纬向接回29厘米，经向接回85厘米
术：印花，梭织提花
势：2013/2014国际流行趋势之艺术家表现形态
器：服装面料，装饰布

Name: Palace Pattern of Ink: Realization of the Dream (1)
Conviction: Chinese painting aesthetics, jazz
Principle: ink shading, rhyme
Method: 4 sides in continuation, zonal tying-in 29 cm, longitudinal tying-in 85cm
Operation: printing, jacquard weaving
Trend: 2013/2014 international fashion trend of artists expression
Application: clothing fabrics, decorative cloth

Creative Fabrics Design 1

名：本杰明的牙买加（1）
道：中国画审美观、爵士乐
理：水墨晕、韵
法：四方连缀，纬向接回29厘米，经向接回85厘米
术：印花，梭织提花
势：2013/2014国际流行趋势之艺术家表现形态
器：服装面料、装饰布

Name: enjamin of Jamaica (1)
Conviction: Chinese painting aesthetics, jazz
Principle: ink shading, rhyme
Method: 4 sides in continuation, zonal tying-in 29 cm, longitudinal tying-in 85cm
Operation: printing, jacquard weaving
Trend: 2013/2014 international fashion trend of artists expression
Application: clothing fabrics, decorative cloth

Creative Fabrics Design 1

名：陈哥辛之恋情意象次序（1）（局部）、花香袭人意象次序（1）（局部）
道：中国画审美观、爵士乐
理：水墨晕、韵
法：四方连缀、纬向接回29厘米、经向接回29厘米
术：印花、梭织提花
势：2013/2014国际流行趋势之艺术家表现形态
器：服装面料、装饰布

Name: Image Sequence of Chen Gexinn's Romance (portion), Image Sequence of Fragrant Flowers (1) (portion)
Conviction: Chinese painting aesthetics, jazz
Principle: ink shading, rhyme
Method: 4 sides in continuation, zonal tying-in 29 cm, longitudinal tying-in 29cm
Operation: printing, jacquard weaving
Trend: 2013/2014 international fashion trend of artists expression
Application: clothing fabrics, decorative cloth

Creative Fabrics Design 1

名：花香袭人意象次序（3~20）
道：中国画审美
理：交互，无序到有序
法：四方连缀、纬向接回29厘米，经向接回29厘米
术：梭织提花、轧花、涂层
势：2013/2014国际流行趋势之艺术家表现形态
器：女装面料

Name: Image Sequence of Fragrant Flowers (3~20)
Conviction: Chinese painting aesthetic
Principle: interaction, from disorderly to orderly
Method: 4 sides in continuation, zonal tying-in 29 cm, longitudinal tying-in 29cm
Operation: jacquard weaving, embossing, coating
Trend: 2013/2014 international fashion trend of artists expression
Application: women's fabric

Creative Fabrics Design 1

名：正是千年忆往事（2）、（31）
道：民俗
理：银蚀刻、表面颗粒或绒质
法：二方连缀、纬向接回35厘米、经向定位85厘米
术：梭织提花、剪花、烂花、烧毛、植绒
势：2013/2014国际流行趋势之自然地貌形态
器：服装面料、装饰布

Name: Long Time Memory of the Past (2), (31)
Conviction: folk
Principle: silver etching, surface particles or fleece
Method: 2 sides in continuation, zonal tying-in 35 cm, longitudinal positioning 85 cm
Operation: jacquard weaving, carving, burnt-discharging, singeing, flocking
Trend: 2013/2014 international fashion trend of natural landform
Application: clothing fabrics, decorative cloth

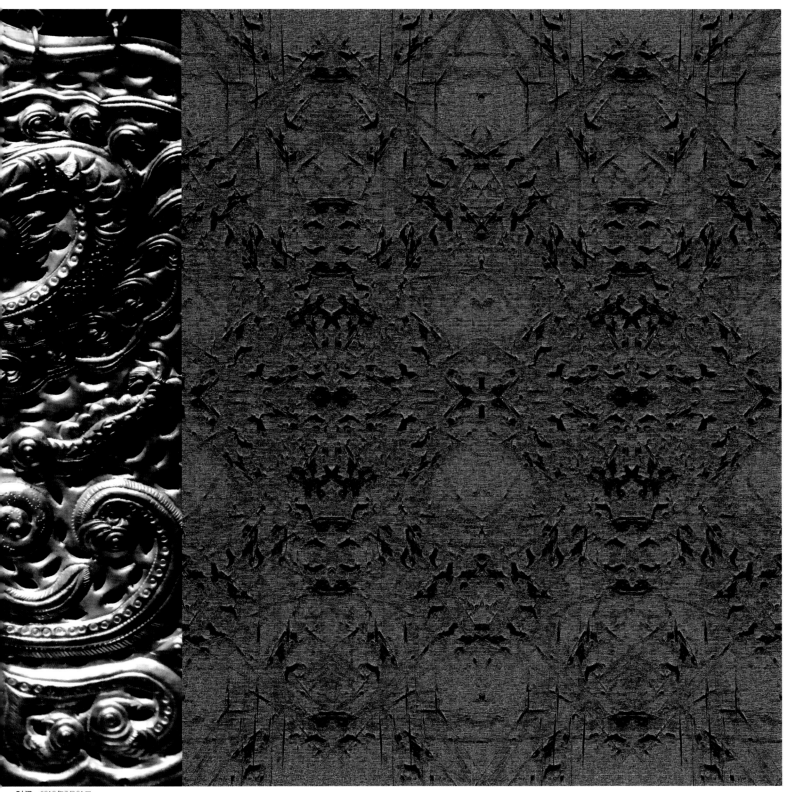

时间：2010年7月31日
地点：四川成都四川博物院
对象：苗族服饰
灵感：图腾威势
Time: July 31, 2010
Location: Sichuan Museum, Chengdu, Sichuan
Object: the Miao costumes
Inspiration: totem art

Creative Fabrics Design 1

名：正是千年忆往事（23）、（24）
道：民俗
理：银蚀刻、表面颗粒或绒质
法：二方连缀、纬向接回35厘米、经向定位85厘米
术：梭织提花、剪花、烂花、烧毛、植绒
势：2013/2014国际流行趋势之自然地貌形态
器：服装面料、装饰布

Name: Long Time Memory of the Past (23), (24)
Conviction: folk
Principle: silver etching, surface particles or fleece
Method: 2 sides in continuation, zonal tying-in 35 cm, longitudinal positioning 85 cm
Operation: jacquard weaving, carving, burnt-discharging, singeing, flocking
Trend: 2013/2014 international fashion trend of natural landform
Application: clothing fabrics, decorative cloth

时间：2012年8月4日
地点：云南省博物馆
对象：苗族服饰
灵感：图腾威势
Time: August 4, 2012
Location: Yunnan Museum, Kunming, Yunnan
Object: the Miao costumes
Inspiration: totem artz

Creative Fabrics Design 1

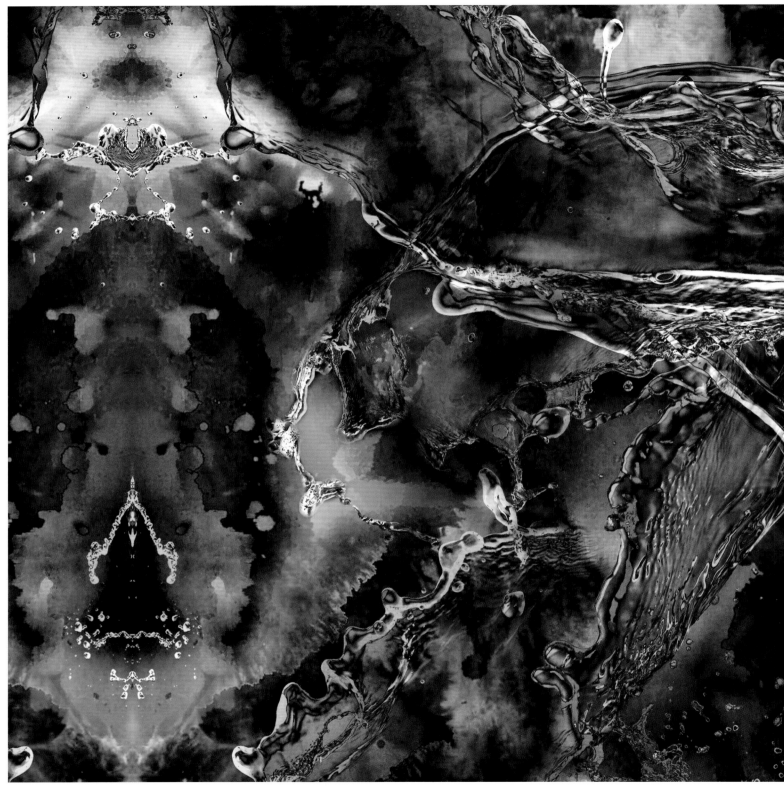

名：小雪节气感恩绵绵（1）（局部）
道：中国画审美观、中国道家五行文化
理：水墨晕、韵
法：四方连缀，纬向接回250厘米，经向接回64.2厘米
术：印花，梭织提花
势：2013/2014国际流行趋势之艺术家表现形态
器：服装面料、装饰布
Name: Light Snow (Solar Term) — Love with Thankfulness (1) (portion)
Conviction: Chinese painting aesthetics, the five elements in Chinese Taoist culture
Principle: ink shading, rhyme
Method: 4 sides in continuation, zonal tying-in 250 cm, longitudinal tying-in 64.2 cm
Operation: printing, jacquard weaving
Trend: 2013/2014 international fashion trend of artists expression
Application: clothing fabrics, decorative cloth

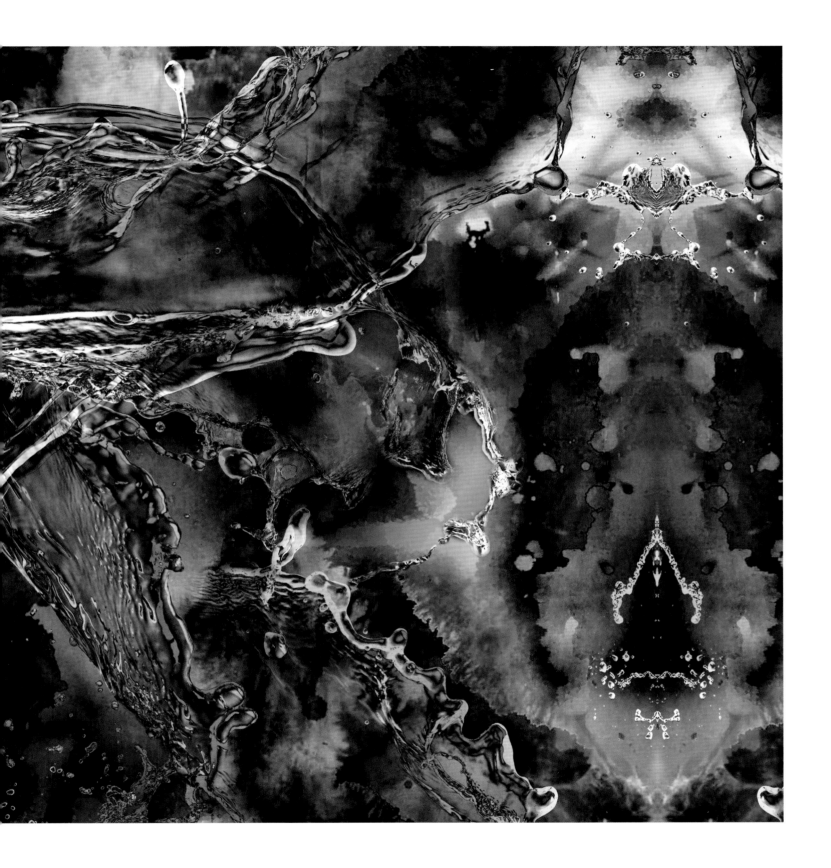

Creative Fabrics Design 1

名：小雪节气感恩绵绵（2）（局部）
道：中国画审美观、中国道家五行文化；
理：水墨晕、韵
法：四方连缀，纬向接回250厘米，经向接回64.2厘米
术：印花，梭织提花
势：2013/2014国际流行趋势之艺术家表现形态
器：服装面料、装饰布

Name: Light Snow (Solar Term) — Love with Thankfulness (2) (protion)
Conviction: Chinese painting aesthetics, the five elements in Chinese Taoist culture
Principle: ink shading, rhyme
Method: 4 sides in continuation, zonal tying-in 250 cm, longitudinal tying-in 64.2 cm
Operation: printing, jacquard weaving
Trend: 2013/2014 international fashion trend of artists expression
Application: clothing fabrics, decorative cloth

Creative Fabrics Design 1

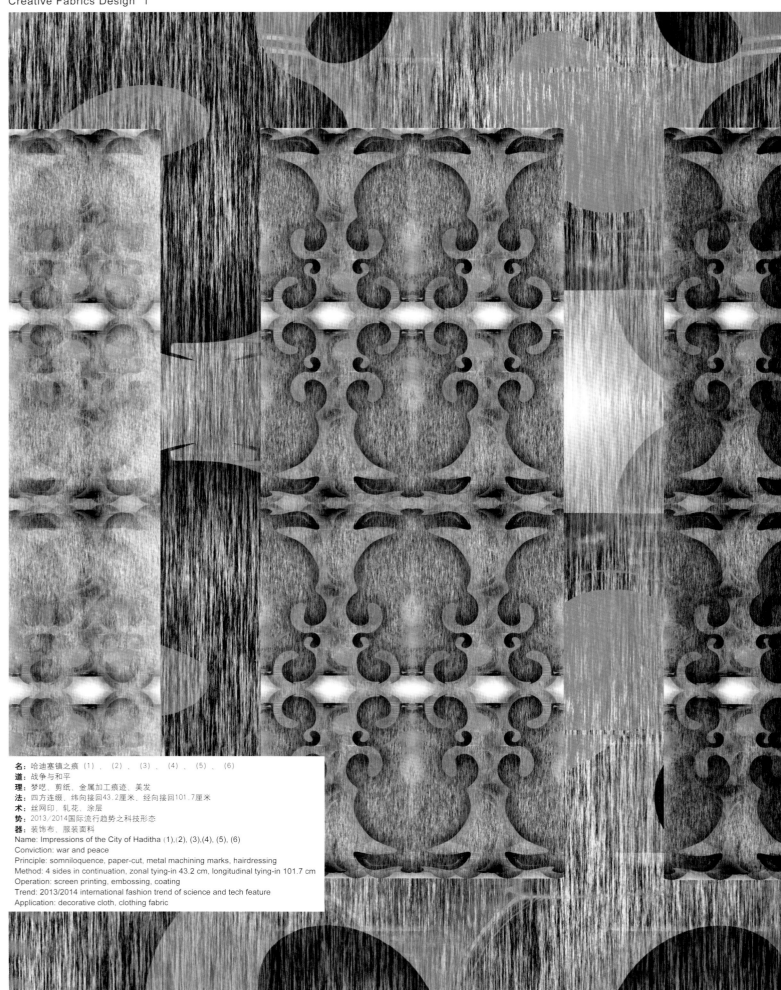

名：哈迪塞镇之痕（1）、（2）、（3）、（4）、（5）、（6）
道：战争与和平
理：梦呓、剪纸、金属加工痕迹、美发
法：四方连缀，纬向接回43.2厘米、经向接回101.7厘米
术：丝网印，轧花、涂层
势：2013/2014国际流行趋势之科技形态
器：装饰布、服装面料

Name: Impressions of the City of Haditha (1),(2), (3),(4), (5), (6)
Conviction: war and peace
Principle: somniloquence, paper-cut, metal machining marks, hairdressing
Method: 4 sides in continuation, zonal tying-in 43.2 cm, longitudinal tying-in 101.7 cm
Operation: screen printing, embossing, coating
Trend: 2013/2014 international fashion trend of science and tech feature
Application: decorative cloth, clothing fabric

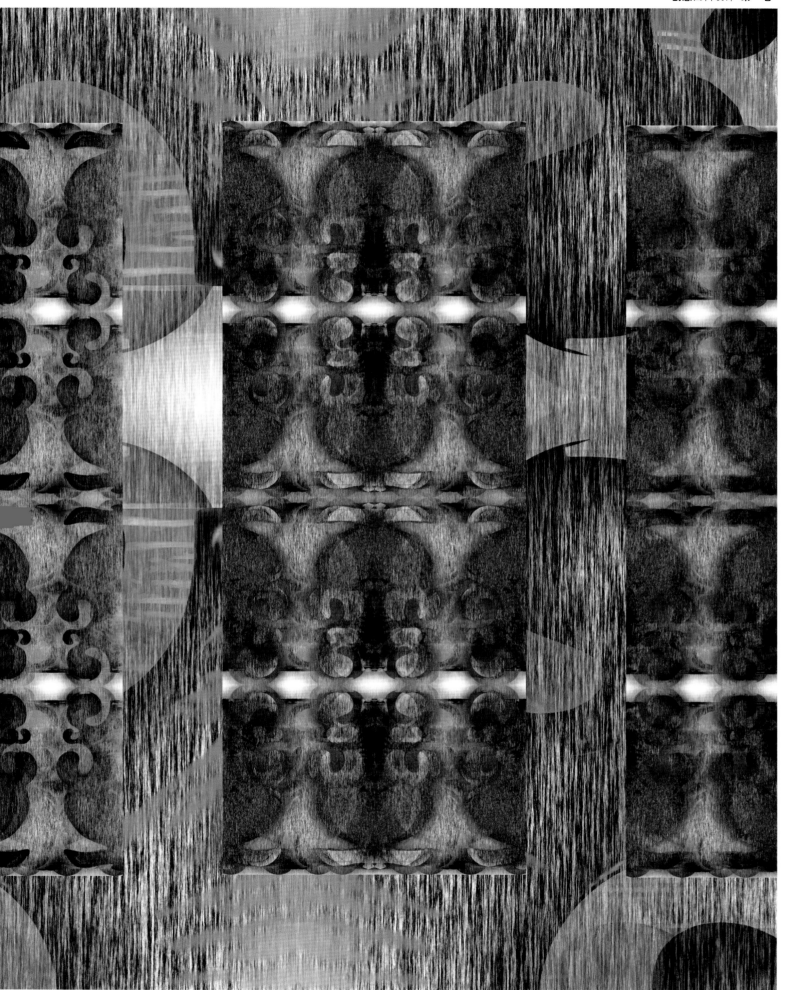

Creative Fabrics Design 1

名：魅影重重——牡丹（1）、（2）
道：中国水墨画审美、中国道家五行文化
理：墨晕、韵
法：二方连缀、纬向接回60厘米、经向定位140厘米
术：梭织提花、转移印花
势：2013/2014国际流行趋势之怀旧文化形态
器：服装面料

Name: Phantomlike — Peony (1), (2)
Conviction: Chinese ink painting aesthetics, the five elements in Chinese Taoist culture
Principle: ink shading, rhyme
Method: 2 sides in continuation, zonal tying-in 60 cm, longitudinal positioning 140 cm
Operation: jacquard weaving, transfer printing
Trend: 2013/2014 international fashion trend of nostalgia
Application: clothing fabrics

时间：2011年12月22日
地点：浴缸
对象：水墨
灵感：水与墨的流动变化
Time: December 22, 2011
Location: the bathtub
Object: ink
Inspiration: changes in the flow of water and ink

Creative Fabrics Design 1

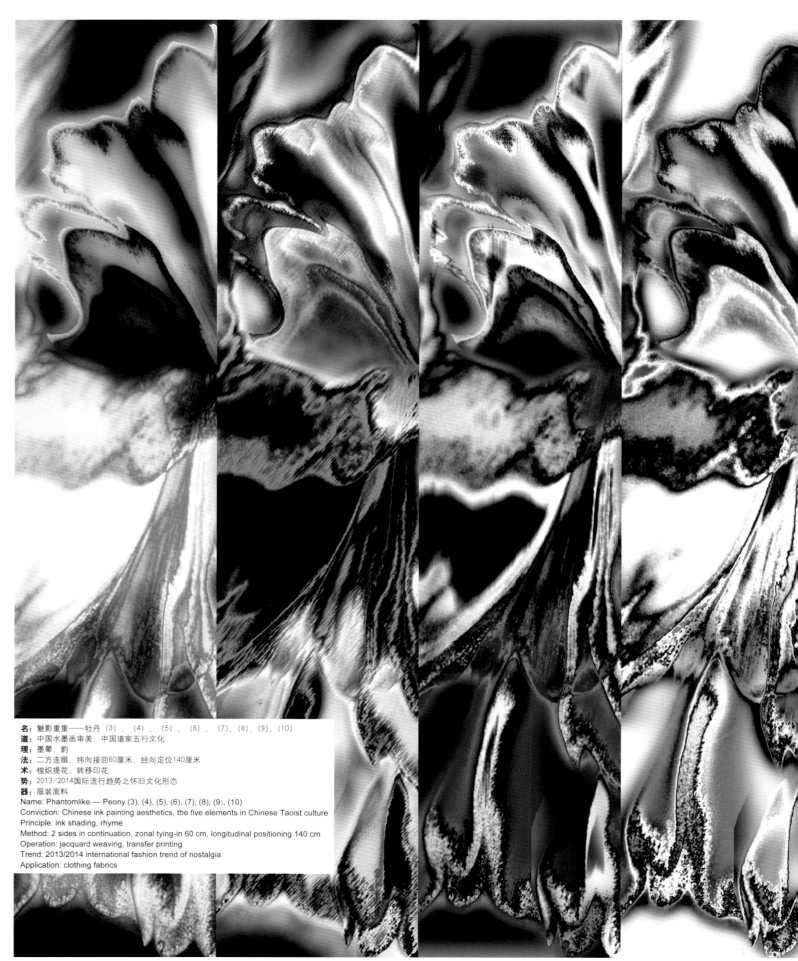

名：魅影重重——牡丹（3）、（4）、（5）、（6）、（7）、（8）、（9）、（10）
道：中国水墨画审美，中国道家五行文化
理：墨晕、韵
法：二方连缀、纬向接回60厘米，经向定位140厘米
术：梭织提花、转移印花
势：2013/2014国际流行趋势之怀旧文化形态
器：服装面料

Name: Phantomlike — Peony (3), (4), (5), (6), (7), (8), (9), (10)
Conviction: Chinese ink painting aesthetics, the five elements in Chinese Taoist culture
Principle: ink shading, rhyme
Method: 2 sides in continuation, zonal tying-in 60 cm, longitudinal positioning 140 cm
Operation: jacquard weaving, transfer printing
Trend: 2013/2014 international fashion trend of nostalgia
Application: clothing fabrics

Creative Fabrics Design 1

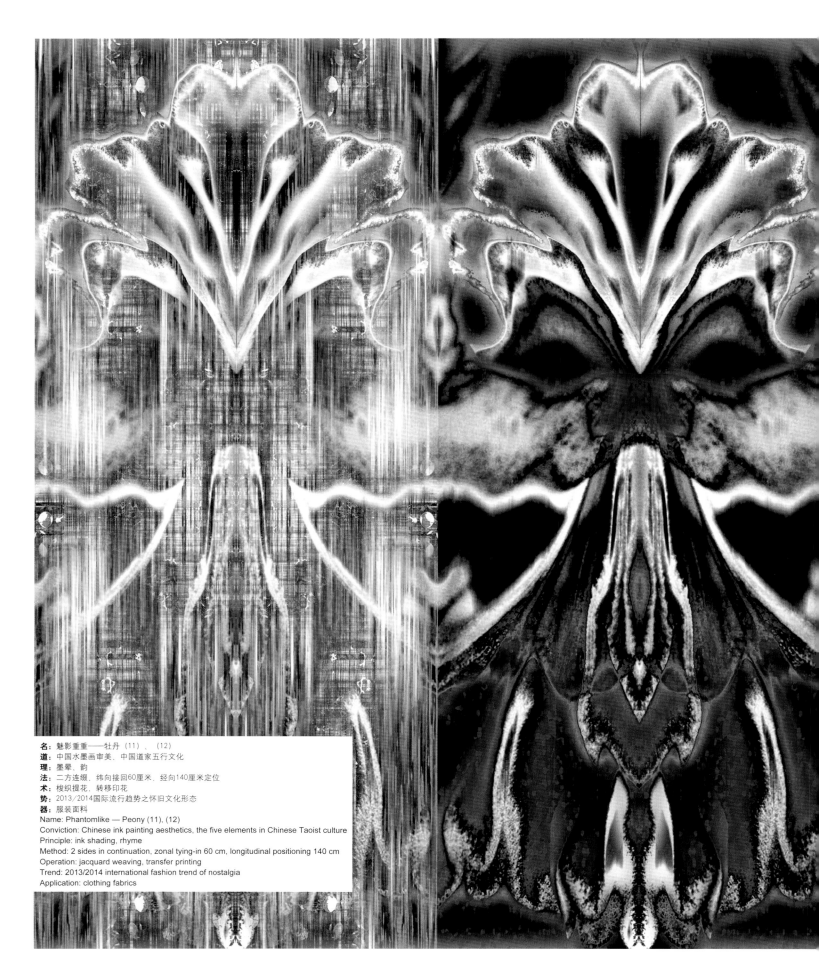

名：魅影重重——牡丹（11）、（12）
道：中国水墨画审美、中国道家五行文化
理：墨晕、韵
法：二方连缀、纬向接回60厘米，经向140厘米定位
术：梭织提花、转移印花
势：2013/2014国际流行趋势之怀旧文化形态
器：服装面料

Name: Phantomlike — Peony (11), (12)
Conviction: Chinese ink painting aesthetics, the five elements in Chinese Taoist culture
Principle: ink shading, rhyme
Method: 2 sides in continuation, zonal tying-in 60 cm, longitudinal positioning 140 cm
Operation: jacquard weaving, transfer printing
Trend: 2013/2014 international fashion trend of nostalgia
Application: clothing fabrics

Creative Fabrics Design 1

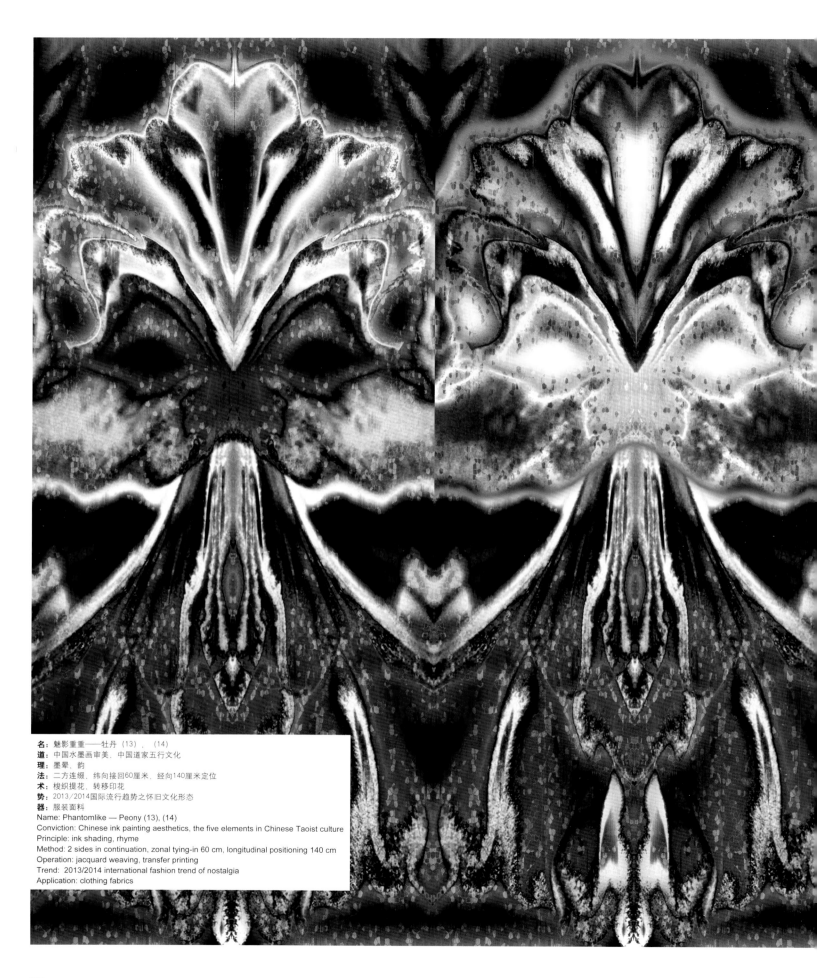

名：魅影重重——牡丹 (13)、(14)
道：中国水墨画审美、中国道家五行文化
理：墨晕、韵
法：二方连续、纬向接回60厘米，经向140厘米定位
术：梭织提花、转移印花
势：2013/2014国际流行趋势之怀旧文化形态
器：服装面料

Name: Phantomlike — Peony (13), (14)
Conviction: Chinese ink painting aesthetics, the five elements in Chinese Taoist culture
Principle: ink shading, rhyme
Method: 2 sides in continuation, zonal tying-in 60 cm, longitudinal positioning 140 cm
Operation: jacquard weaving, transfer printing
Trend: 2013/2014 international fashion trend of nostalgia
Application: clothing fabrics

时间：2008年9月5日
地点：东京国立博物馆
对象：面具
灵感：图腾威势
Time: September 5, 2008
Location: National Museum of Japan, Tokyo
Object: the mask
Inspiration: totem art

Creative Fabrics Design 1

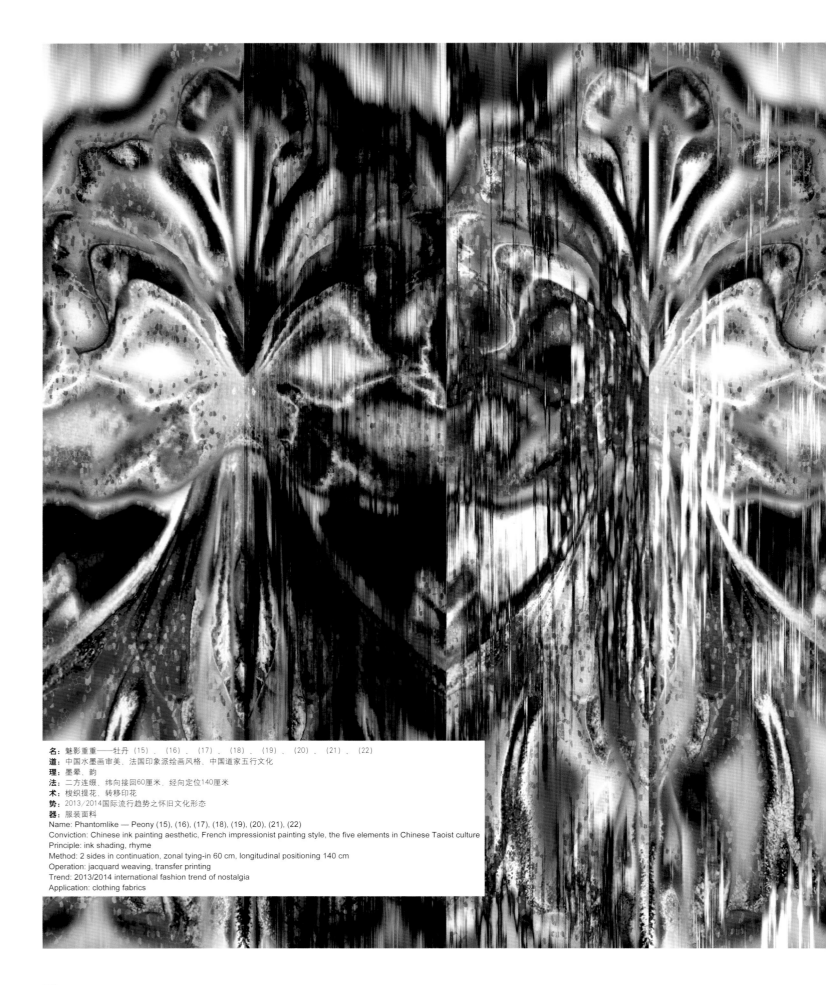

名：魅影重重——牡丹（15）、（16）、（17）、（18）、（19）、（20）、（21）、（22）
道：中国水墨画审美、法国印象派绘画风格、中国道家五行文化
理：墨晕、韵
法：二方连缀、纬向接回60厘米、经向定位140厘米
术：梭织提花、转移印花
势：2013/2014国际流行趋势之怀旧文化形态
器：服装面料
Name: Phantomlike — Peony (15), (16), (17), (18), (19), (20), (21), (22)
Conviction: Chinese ink painting aesthetic, French impressionist painting style, the five elements in Chinese Taoist culture
Principle: ink shading, rhyme
Method: 2 sides in continuation, zonal tying-in 60 cm, longitudinal positioning 140 cm
Operation: jacquard weaving, transfer printing
Trend: 2013/2014 international fashion trend of nostalgia
Application: clothing fabrics

Creative Fabrics Design 1

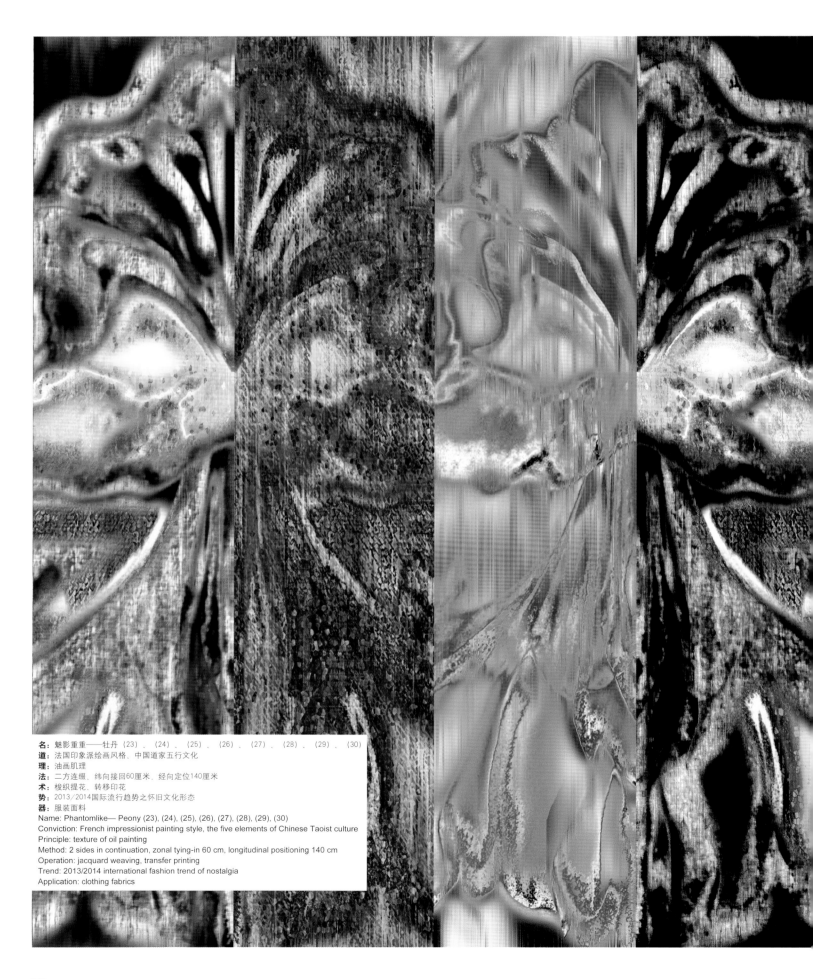

名：魅影重重——牡丹（23）、（24）、（25）、（26）、（27）、（28）、（29）、（30）
道：法国印象派绘画风格、中国道家五行文化
理：油画肌理
法：二方连缀，纬向接回60厘米，经向定位140厘米
术：梭织提花、转移印花
势：2013/2014国际流行趋势之怀旧文化形态
器：服装面料

Name: Phantomlike— Peony (23), (24), (25), (26), (27), (28), (29), (30)
Conviction: French impressionist painting style, the five elements of Chinese Taoist culture
Principle: texture of oil painting
Method: 2 sides in continuation, zonal tying-in 60 cm, longitudinal positioning 140 cm
Operation: jacquard weaving, transfer printing
Trend: 2013/2014 international fashion trend of nostalgia
Application: clothing fabrics

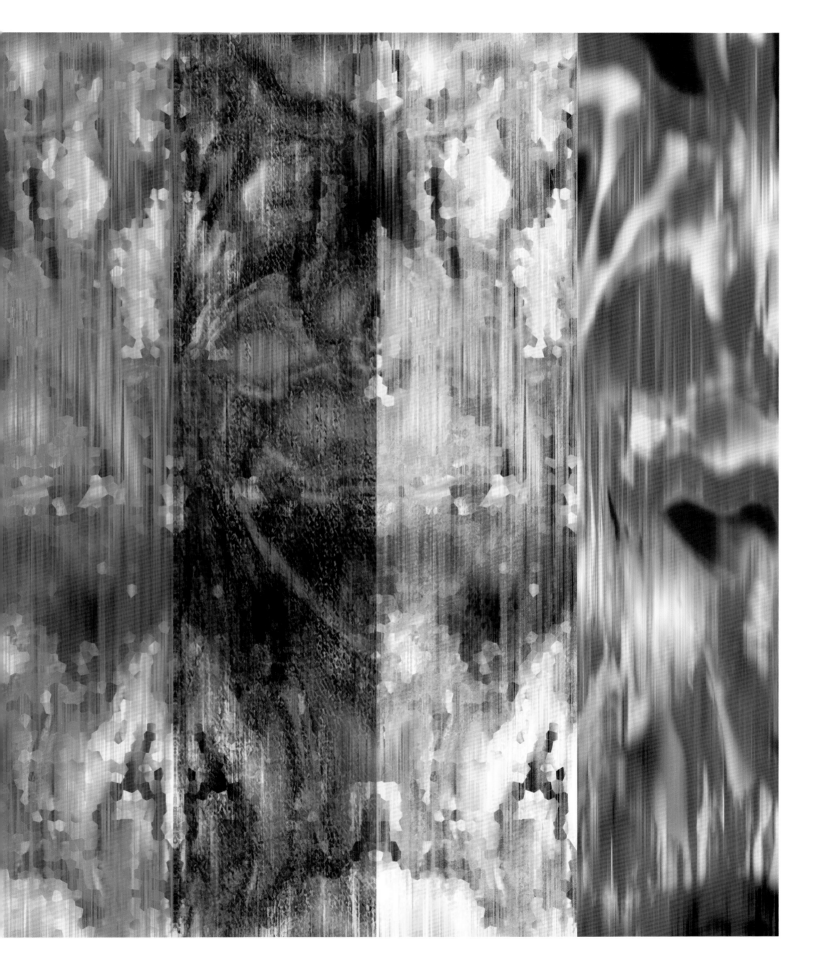

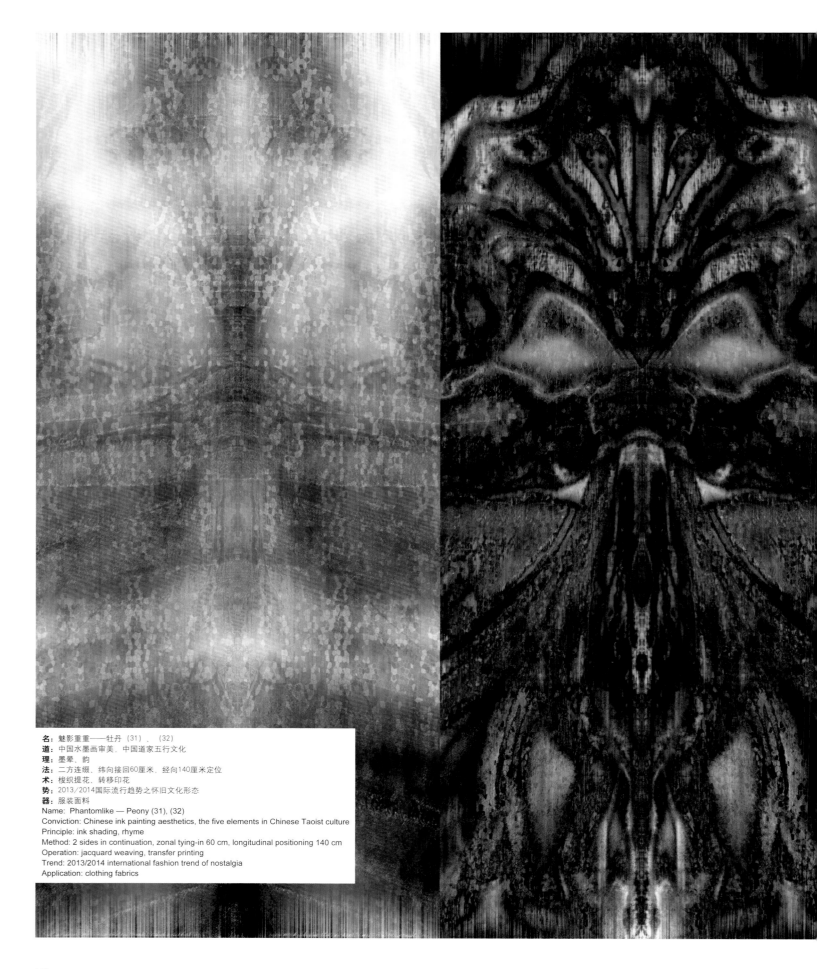

名：魅影重重——牡丹（31）、(32)
道：中国水墨画审美，中国道家五行文化
理：墨晕、韵
法：二方连缀、纬向接回60厘米，经向140厘米定位
术：梭织提花、转移印花
势：2013/2014国际流行趋势之怀旧文化形态
器：服装面料
Name: Phantomlike — Peony (31), (32)
Conviction: Chinese ink painting aesthetics, the five elements in Chinese Taoist culture
Principle: ink shading, rhyme
Method: 2 sides in continuation, zonal tying-in 60 cm, longitudinal positioning 140 cm
Operation: jacquard weaving, transfer printing
Trend: 2013/2014 international fashion trend of nostalgia
Application: clothing fabrics

Creative Fabrics Design 1

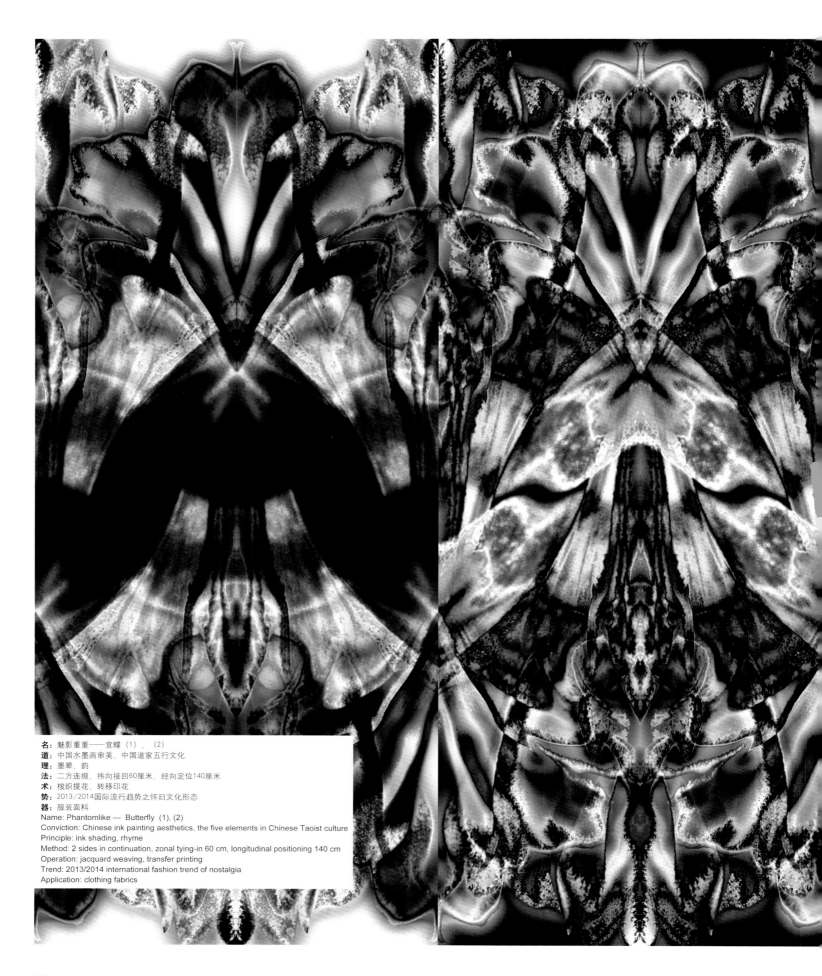

名：魅影重重——宣蝶（1）、（2）
道：中国水墨画审美、中国道家五行文化
理：墨晕、韵
法：二方连缀、纬向接回60厘米，经向定位140厘米
术：梭织提花、转移印花
势：2013/2014国际流行趋势之怀旧文化形态
器：服装面料

Name: Phantomlike — Butterfly (1), (2)
Conviction: Chinese ink painting aesthetics, the five elements in Chinese Taoist culture
Principle: ink shading, rhyme
Method: 2 sides in continuation, zonal tying-in 60 cm, longitudinal positioning 140 cm
Operation: jacquard weaving, transfer printing
Trend: 2013/2014 international fashion trend of nostalgia
Application: clothing fabrics

时间：2008年10月31日
地点：北京奥林匹克中心
对象：风景镜像
灵感：幻影
Time: October 31, 2008
Location: Beijing Olympic Center
Object: landscape image
Inspiration: phantom

Creative Fabrics Design 1

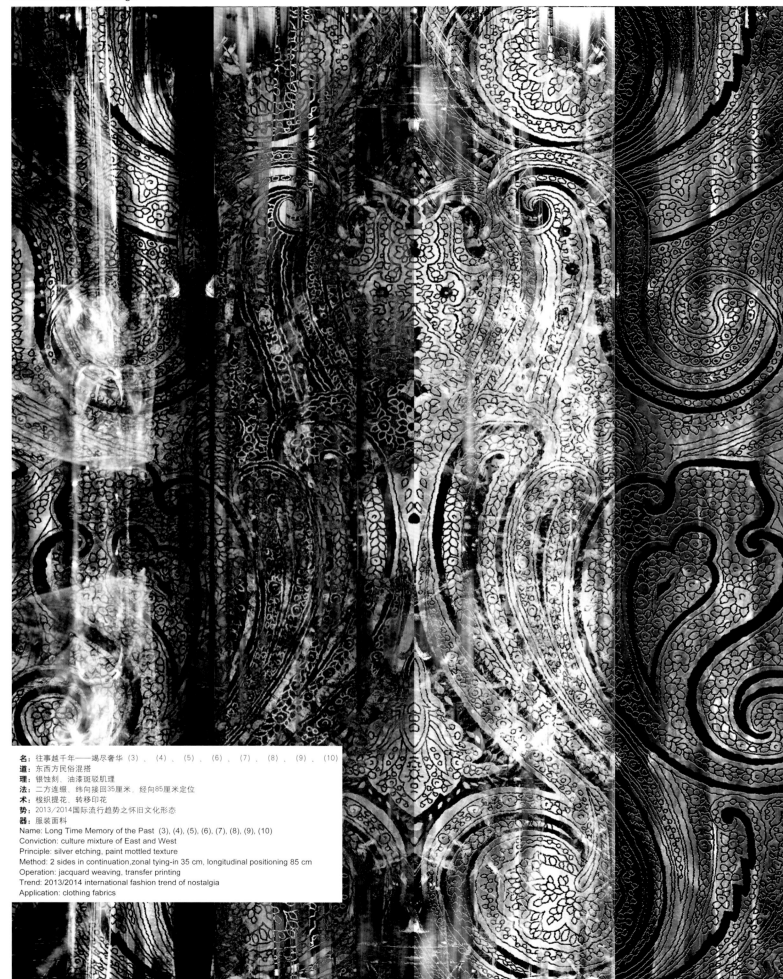

名：往事越千年——竭尽奢华（3）、（4）、（5）、（6）、（7）、（8）、（9）、（10）
道：东西方民俗混搭
理：银蚀刻、油漆斑驳肌理
法：二方连缀，纬向接回35厘米，经向85厘米定位
术：梭织提花、转移印花
势：2013/2014国际流行趋势之怀旧文化形态
器：服装面料

Name: Long Time Memory of the Past (3), (4), (5), (6), (7), (8), (9), (10)
Conviction: culture mixture of East and West
Principle: silver etching, paint mottled texture
Method: 2 sides in continuation, zonal tying-in 35 cm, longitudinal positioning 85 cm
Operation: jacquard weaving, transfer printing
Trend: 2013/2014 international fashion trend of nostalgia
Application: clothing fabrics

Creative Fabrics Design 1

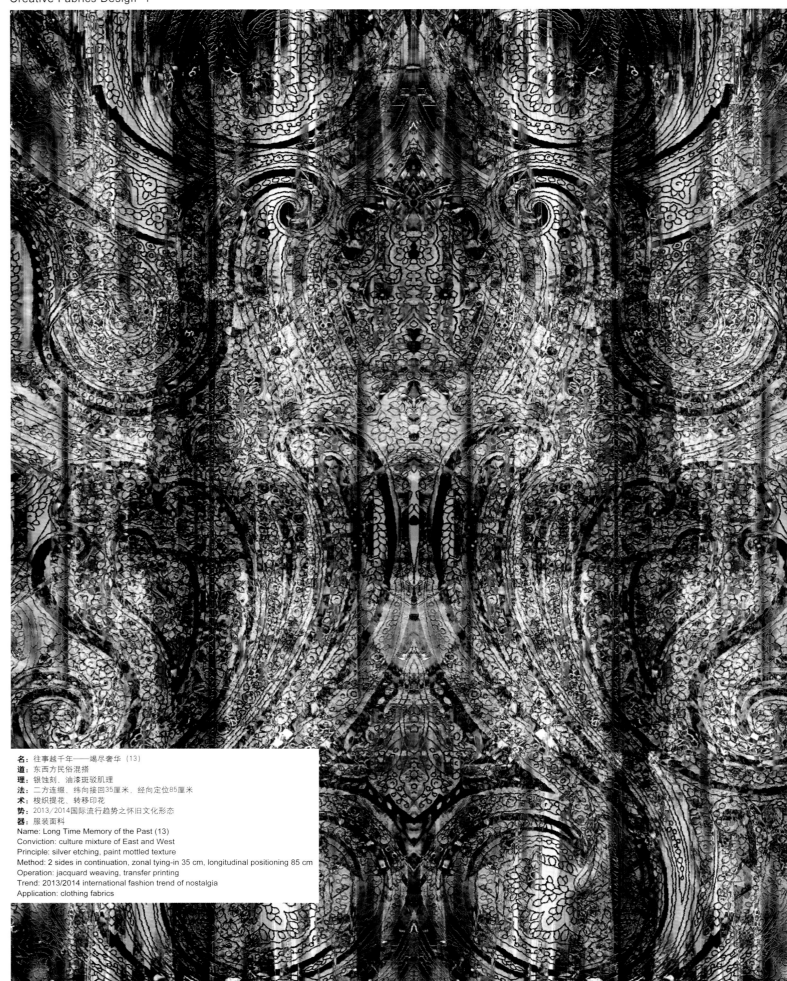

名：往事越千年——竭尽奢华（13）
道：东西方民俗混搭
理：银蚀刻、油漆斑驳肌理
法：二方连缀、纬向接回35厘米、经向定位85厘米
术：梭织提花、转移印花
势：2013/2014国际流行趋势之怀旧文化形态
器：服装面料

Name: Long Time Memory of the Past (13)
Conviction: culture mixture of East and West
Principle: silver etching, paint mottled texture
Method: 2 sides in continuation, zonal tying-in 35 cm, longitudinal positioning 85 cm
Operation: jacquard weaving, transfer printing
Trend: 2013/2014 international fashion trend of nostalgia
Application: clothing fabrics

时间：2012年7月31日
地点：四川成都四川博物院
对象：皮影
灵感：木刻浮雕
Time: July 31, 2012
Location: Sichuan Museum, Chengdu, Sichuan
Object: shadow puppet
Inspiration: transparency

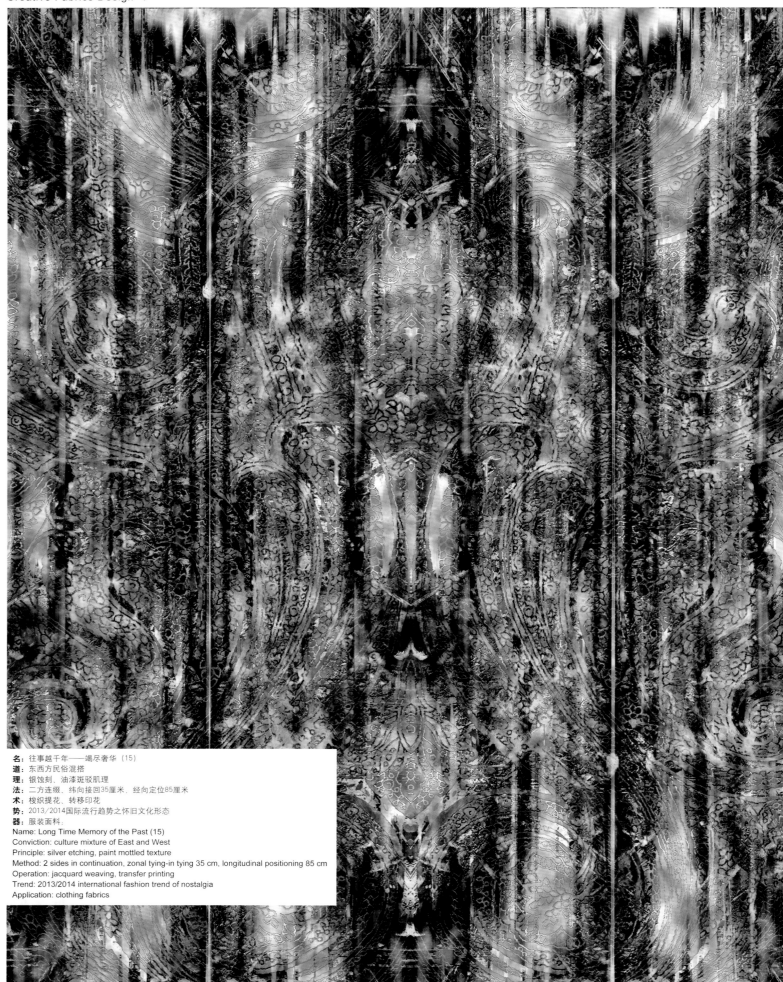

名：往事越千年——竭尽奢华（15）
道：东西方民俗混搭
理：银蚀刻、油漆斑驳肌理
法：二方连缀、纬向接回35厘米，经向定位85厘米
术：梭织提花、转移印花
势：2013/2014国际流行趋势之怀旧文化形态
器：服装面料

Name: Long Time Memory of the Past (15)
Conviction: culture mixture of East and West
Principle: silver etching, paint mottled texture
Method: 2 sides in continuation, zonal tying-in tying 35 cm, longitudinal positioning 85 cm
Operation: jacquard weaving, transfer printing
Trend: 2013/2014 international fashion trend of nostalgia
Application: clothing fabrics

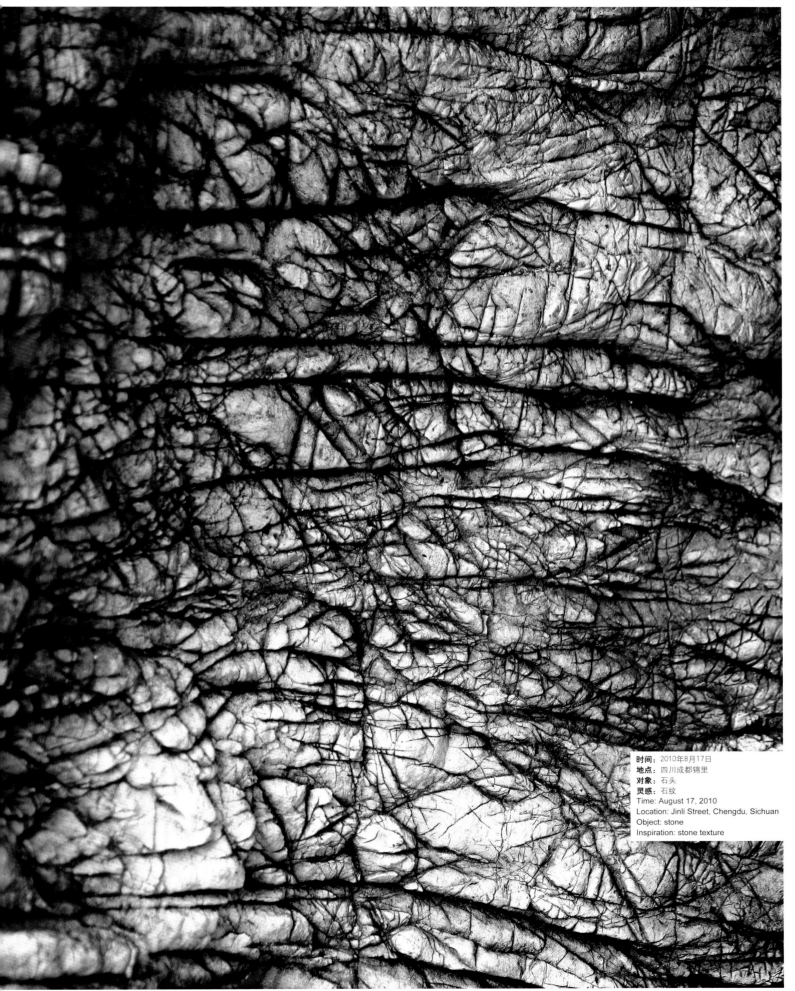

时间：2010年8月17日
地点：四川成都锦里
对象：石头
灵感：石纹
Time: August 17, 2010
Location: Jinli Street, Chengdu, Sichuan
Object: stone
Inspiration: stone texture

Creative Fabrics Design 1

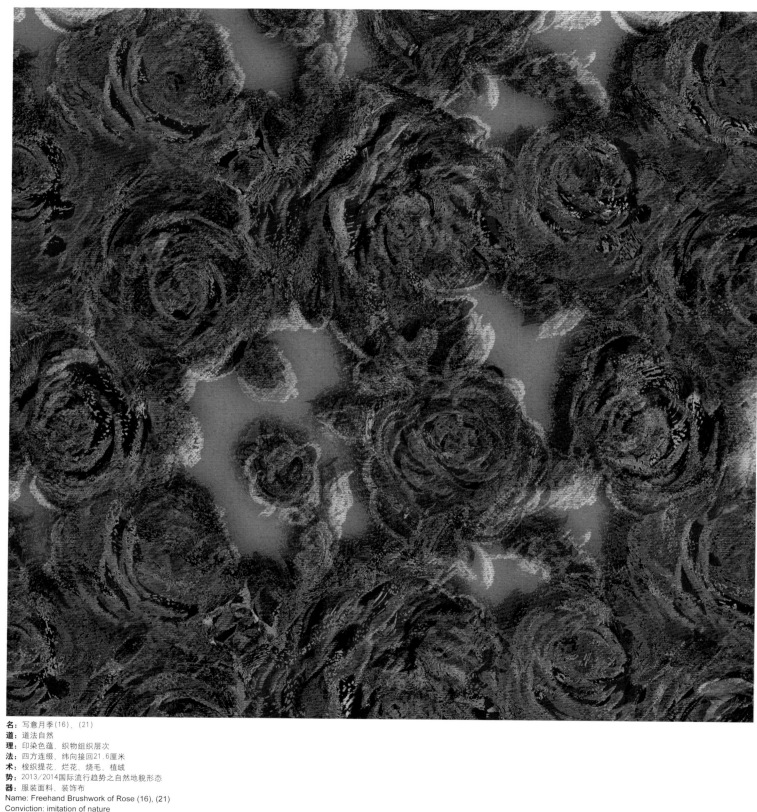

名： 写意月季(16)、(21)
道： 道法自然
理： 印染色蕴、织物组织层次
法： 四方连续、纬向接回21.6厘米
术： 梭织提花、烂花、烧毛、植绒
势： 2013/2014国际流行趋势之自然地貌形态
器： 服装面料、装饰布

Name: Freehand Brushwork of Rose (16), (21)
Conviction: imitation of nature
Principle: printing and dyeing color, fabric texture shade
Method: 4 sides in continuation, zonal tying-in 21.6 cm
Operation: jacquard weaving, burnt-discharging, singeing, flocking
Trend: 2013/2014 international fashion trend of natural landform
Application: clothing fabrics, decorative cloth

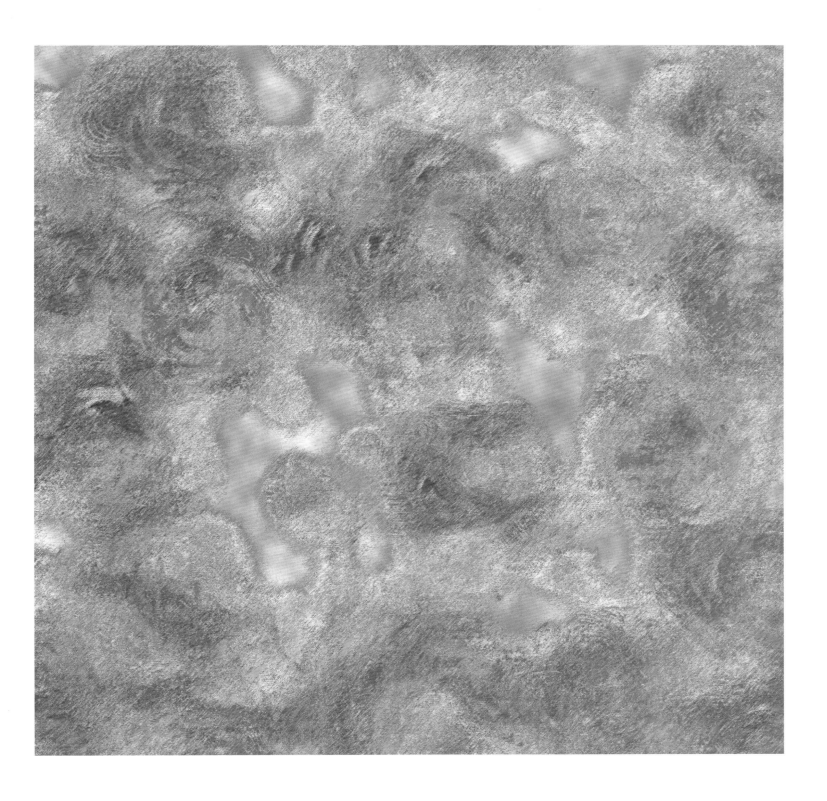

Creative Fabrics Design 1

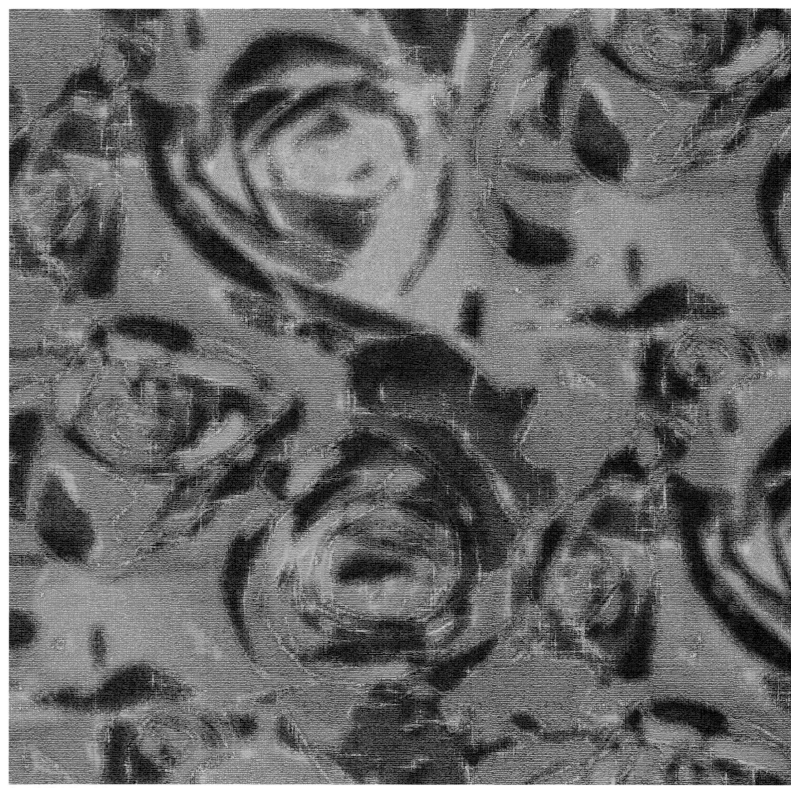

名：月季逐艳（17）、（18）
道：道法自然
理：印染色蕴、织物组织层次
法：四方连缀、纬向接回21.6厘米、经向接回32.1厘米
术：梭织提花、烂花、烧毛、植绒
势：2013/2014国际流行趋势之自然地貌形态
器：服装面料、装饰布

Name: Freehand Brushwork of Rose (16), (21)
Conviction: imitation of nature
Principle: printing and dyeing color, fabric texture shade
Method: 4 sides in continuation, zonal tying-in 21.6 cm
Operation: jacquard weaving, burnt-discharging, singeing, flocking
Trend: 2013/2014 international fashion trend of natural landform
Application: clothing fabrics, decorative cloth

Creative Fabrics Design 1

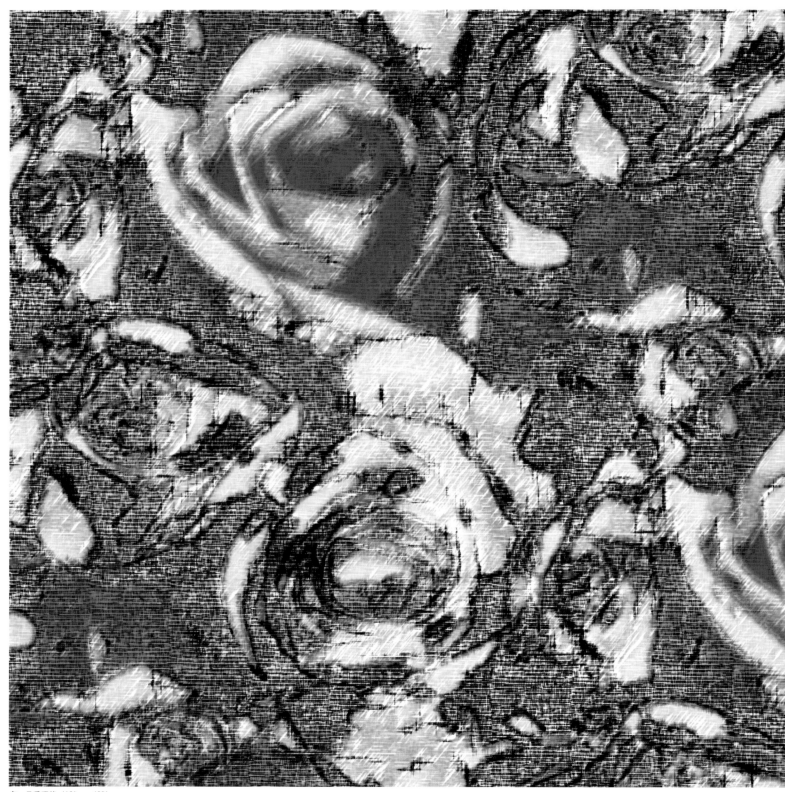

名： 月季逐艳 (19)、(28)
道： 道法自然
理： 印染色蕴，织物组织表现蛇纹层次
法： 四方连缀，纬向接回21.6厘米，经向接回32.1厘米
术： 梭织提花，烂花，烧毛，植绒
势： 2013/2014国际流行趋势之自然生物形态
器： 服装面料、装饰布

Name: Rose in Bloom (19), (28)
Conviction: imitation of nature
Principle: printing and dyeing color, fabric texture of serpentine shade
Method: 4 sides in continuation, zonal tying-in tying 21.6 cm, longitudinal tying-in 32.1 cm
Operation: jacquard weaving, burnt-discharging, singeing, flocking
Trend: 2013/2014 international fashion trend of bio feature
Application: clothing fabrics, decorative cloth

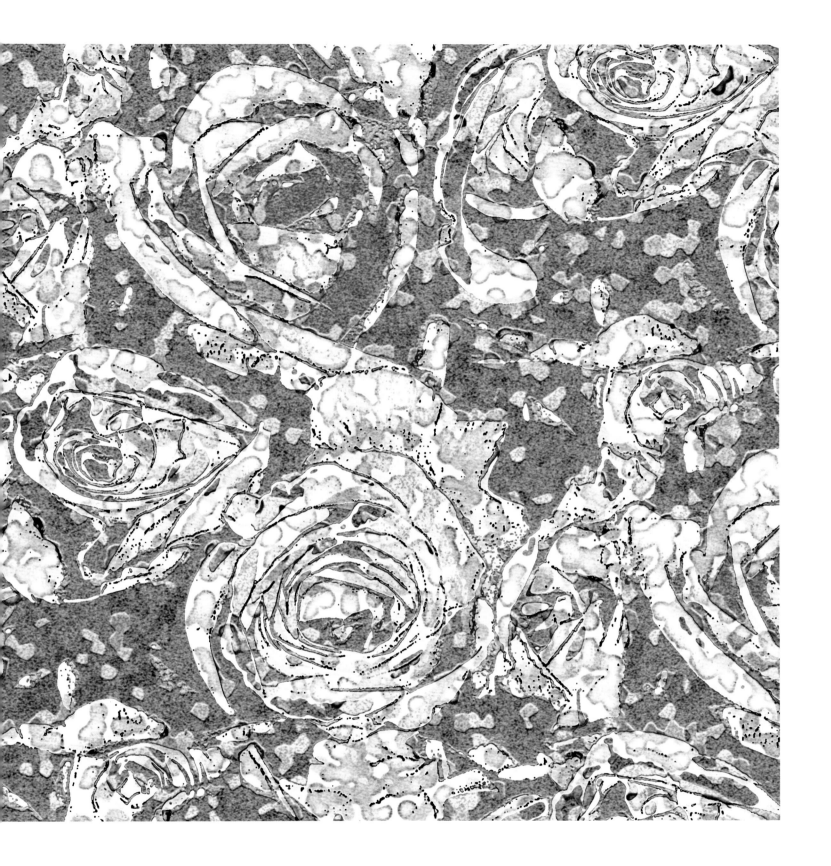

Creative Fabrics Design 1

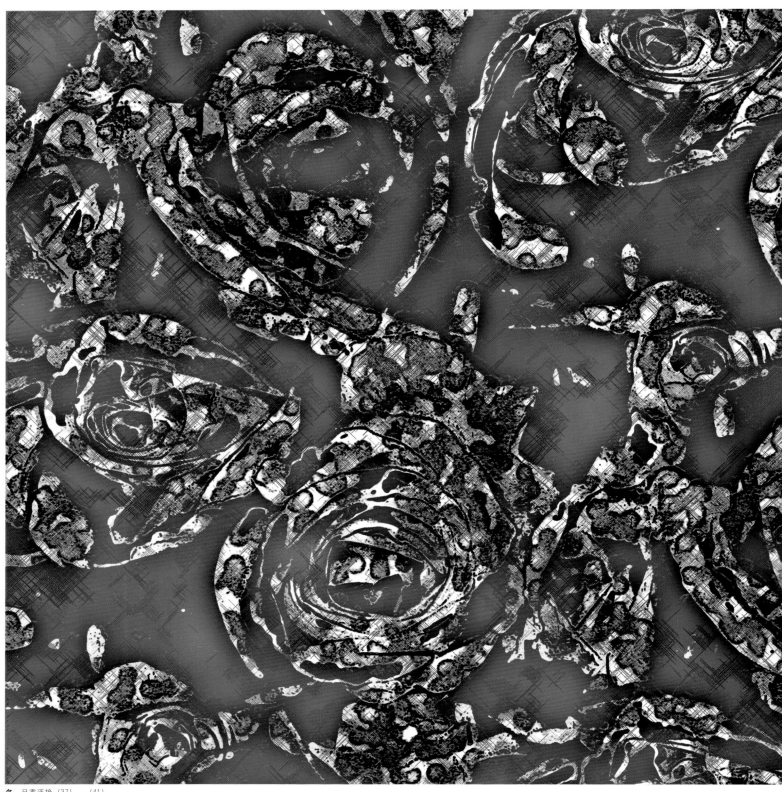

名：月季逐艳（37）、（41）
道：道法自然
理：印染色蕴，织物组织表现蛇纹层次
法：四方连缀，纬向接回21.6厘米，经向接回32.1厘米
术：梭织提花，烂花，轧花，烧毛，植绒
势：2013/2014国际流行趋势之自然生物形态
器：服装面料、装饰布
Name: Rose in Bloom (37), (41)
Conviction: imitation of nature
Principle: printing and dyeing color, fabric texture of serpentine shades
Method: 4 sides in continuation, zonal tying-in 21.6 cm, longitudinal back tying 32.1 cm
Operation: jacquard weaving, burnt-discharging, embossing, singeing, flocking
Trend: 2013/2014 international fashion trend of bio feature
Application: clothing fabrics, decorative cloth

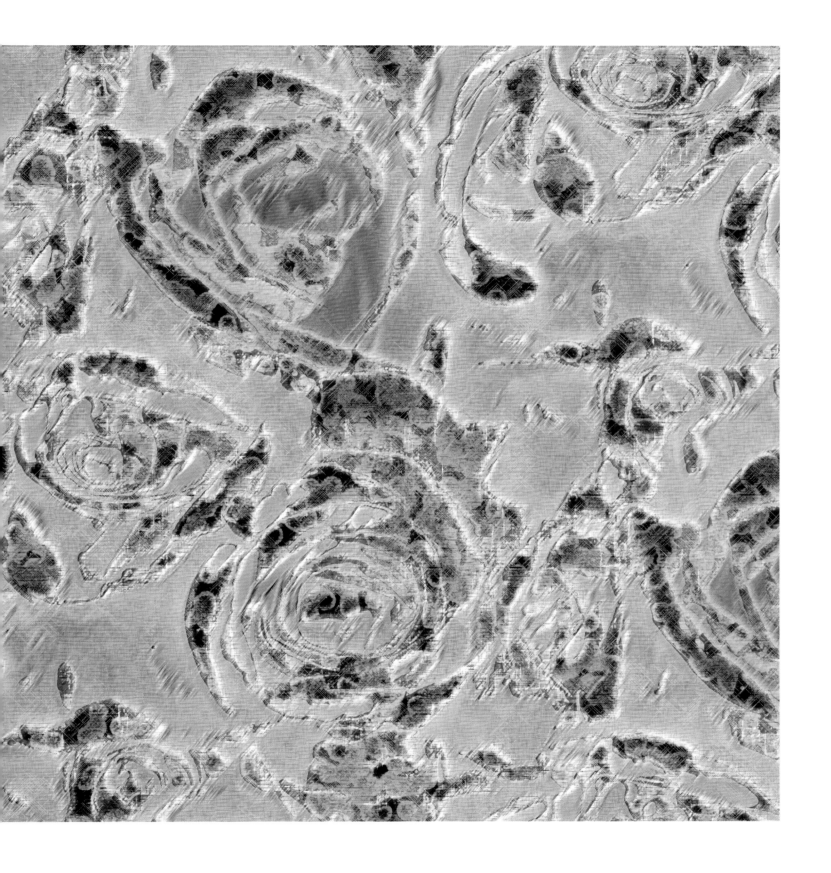

Creative Fabrics Design 1

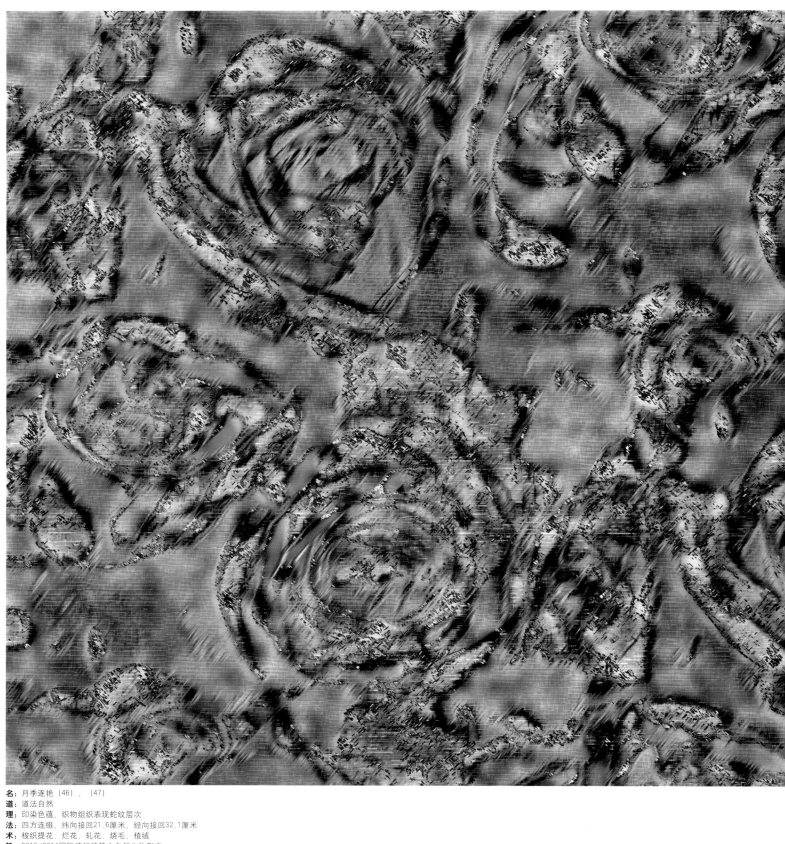

名：月季逐艳（46）、（47）
道：道法自然
理：印染色蕴、织物组织表现蛇纹层次
法：四方连缀、纬向接回21.6厘米、经向接回32.1厘米
术：梭织提花、烂花、轧花、烧毛、植绒
势：2013/2014国际流行趋势之自然生物形态
器：服装面料、装饰布

Name: Rose in Bloom (46), (47)
Conviction: imitation of nature
Principle: printing and dyeing color, fabric texture of serpentine shades
Method: 4 sides in continuation, zonal tying-in 21.6 cm, longitudinal back tying 32.1 cm
Operation: jacquard weaving, burnt-discharging, embossing, singeing, flocking
Trend: 2013/2014 international fashion trend of bio feature
Application: clothing fabrics, decorative cloth

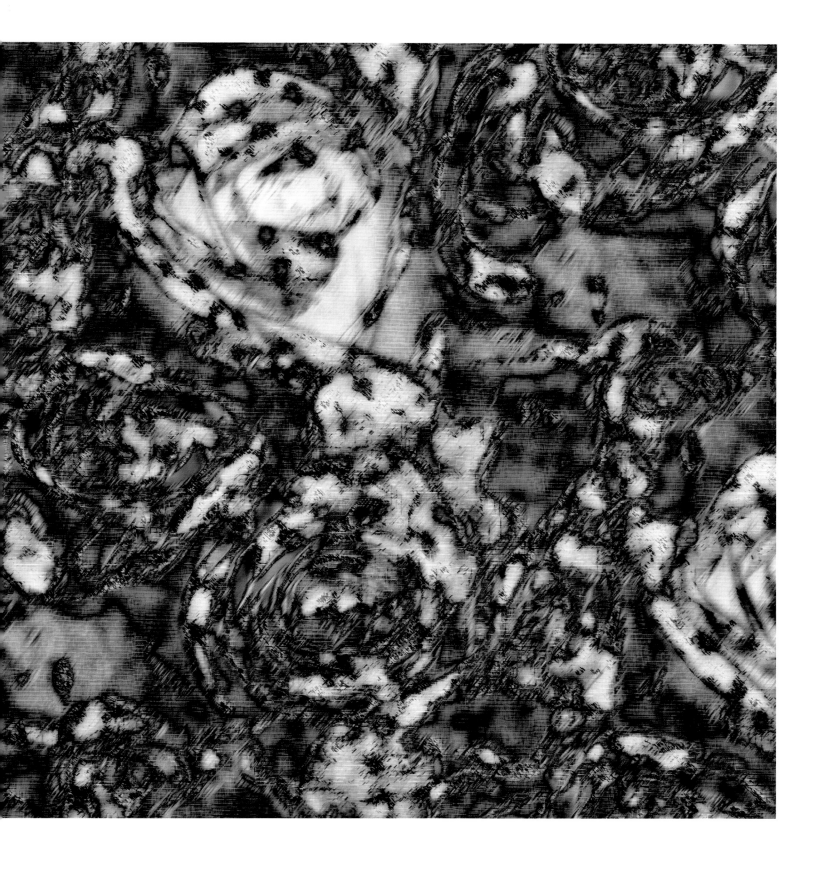

Creative Fabrics Design 1

名：月季再逐艳（14）、（16）
道：道法自然
理：印染色蕴、织物组织层次
法：四方连缀、纬向接回21.6厘米、经向接回32.1厘米
术：梭织提花、烂花、轧花、烧毛、植绒
势：2013/2014国际流行趋势之自然生物形态
器：服装面料、装饰布

Name: Rose in Bloom Once More (14), (16)
Conviction: imitation of nature
Principle: printing and dyeing color, fabric texture
Method: 4 sides in continuation, zonal tying-in 21.6 cm, longitudinal back tying 32.1 cm
Operation: jacquard weaving, burnt-discharging, embossing, crushing
Trend: 2013/2014 international fashion trend of bio feature
Application: clothing fabrics, decorative cloth

Creative Fabrics Design 1

名：月季再逐艳（17）、（18）
道：道法自然
理：印染色蕴、织物组织层次
法：四方连缀、纬向接回21.6厘米、经向接回32.1厘米
术：梭织提花，烂花，轧花，压皱
势：2013/2014国际流行趋势之自然生物形态
器：服装面料、装饰布

Name: Rose and Rose Once More (17), (18)
Conviction: imitation of nature
Principle: printing and dyeing color, fabric texture
Method: 4 sides in continuation, zonal tying-in 21.6 cm, longitudinal tying-in 32.1 cm
Operation: jacquard weaving, burnt-discharge, embossing, crushing
Trend: 2013/2014 international trends of bio feature
Application: clothing fabrics, decorative cloth

Creative Fabrics Design 1

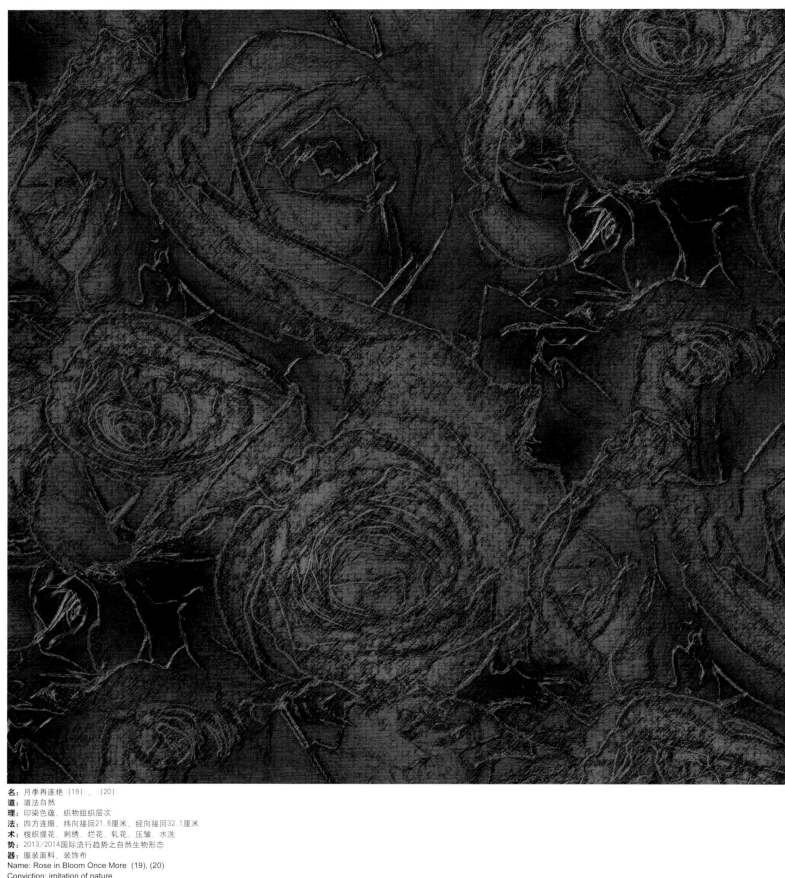

名：月季再逐艳（19）、（20）
道：道法自然
理：印染色蕴、织物组织层次
法：四方连缀、纬向接回21.6厘米、经向接回32.1厘米
术：梭织提花、刺绣、烂花、轧花、压皱、水洗
势：2013/2014国际流行趋势之自然生物形态
器：服装面料、装饰布

Name: Rose in Bloom Once More (19), (20)
Conviction: imitation of nature
Principle: printing and dyeing color, fabric texture
Method: 4 sides in continuation, zonal tying-in 21.6 cm, longitudinal back tying 32.1 cm
Operation: jacquard weaving, embroidery, burnt-discharging, embossing, crushing, washing
Trend: 2013/2014 international fashion trend of bio feature
Application: clothing fabrics, decorative cloth

Creative Fabrics Design 1

名：月季再逐艳（21）、（22）
道：道法自然
理：印染色蕴、织物组织层次
法：四方连缀、纬向接回21.6厘米、经向接回32.1厘米
术：梭织提花、刺绣、水洗、喷染
势：2013/2014国际流行趋势之自然生物形态
器：服装面料、装饰布

Name: Rose in Bloom Once More (21), (22)
Conviction: imitation of nature
Principle: printing and dyeing color, fabric texture
Method: 4 sides in continuation, zonal tying-in 21.6 cm, longitudinal back tying 32.1 cm
Operation: jacquard weaving, embroidery, washing, spraying dye
Trend: 2013/2014 international fashion trend of bio feature
Application: clothing fabrics, decorative cloth

Creative Fabrics Design 1

名：月季再逐艳（23）、（24）
道：道法自然
理：印染色蕴、织物组织层次
法：四方连缀、纬向接回21.6厘米、经向接回32.1厘米
术：梭织提花、刺绣、水洗、喷染
势：2013/2014国际流行趋势之自然生物形态
器：服装面料、装饰布

Name: Rose in Bloom Once More (23), (24)
Conviction: imitation of nature
Principle: printing and dyeing color, fabric texture
Method: 4 sides in continuation, zonal tying-in 21.6 cm, longitudinal back tying 32.1 cm
Operation: jacquard weaving, embroidery, washing, spraying dye
Trend: 2013/2014 international fashion trend of bio feature
Application: clothing fabrics, decorative cloth

Creative Fabrics Design 1

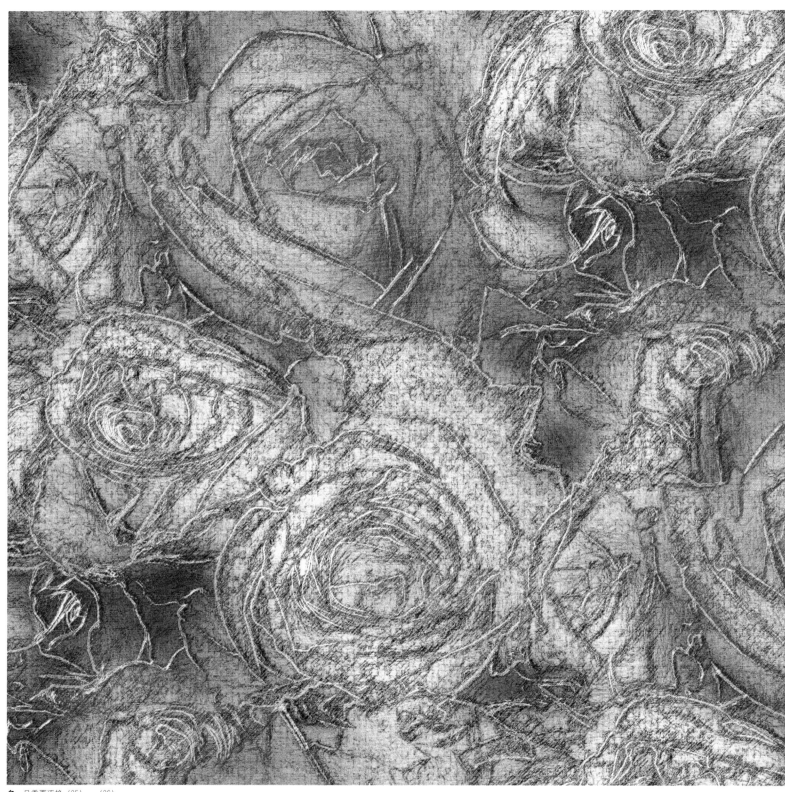

名：月季再逐艳（25）、（26）
道：道法自然
理：印染色蕴、织物组织层次
法：四方连缀、纬向接回21.6厘米、经向接回32.1厘米
术：梭织提花、剪花、刺绣、水洗、喷染
势：2013/2014国际流行趋势之自然生物形态
器：服装面料、装饰布

Name: Rose in Bloom Once More (25), (26)
Conviction: imitation of nature
Principle: printing and dyeing color, fabric texture
Method: 4 sides in continuation, zonal tying-in 21.6 cm, longitudinal back tying, 32.1 cm
Operation: jacquard weaving, cutting flowers, embroidery, washing, spraying dye
Trend: 2013/2014 international fashion trend of bio feature
Application: clothing fabrics, decorative cloth

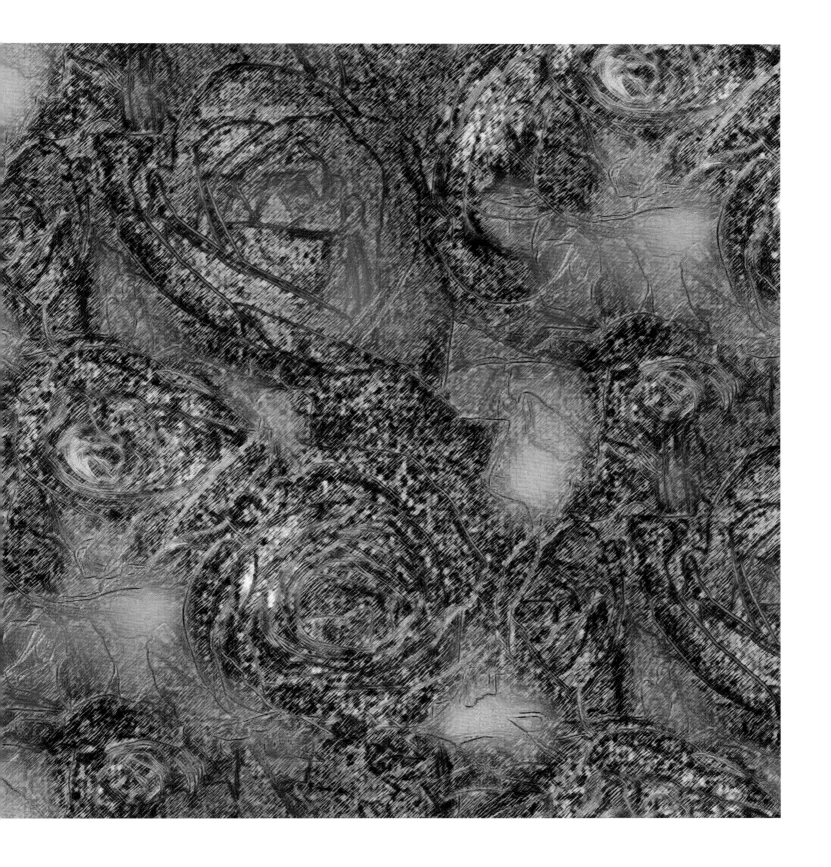

Creative Fabrics Design 1

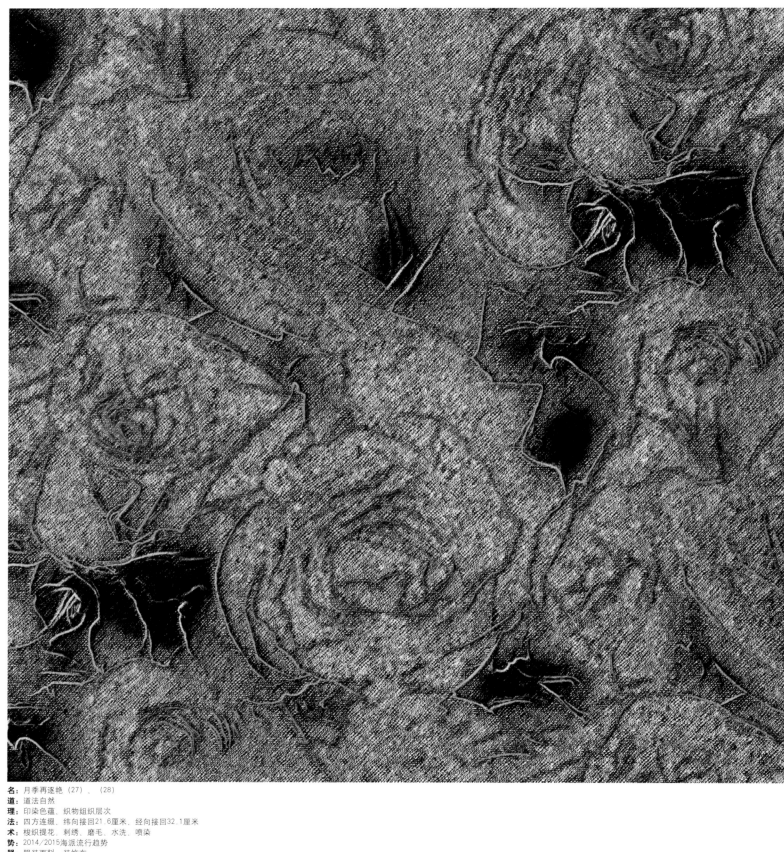

名：月季再逐艳（27）、（28）
道：道法自然
理：印染色蕴、织物组织层次
法：四方连缀、纬向接回21.6厘米、经向接回32.1厘米
术：梭织提花、刺绣、磨毛、水洗、喷染
势：2014/2015海派流行趋势
器：服装面料、装饰布

Name: Rose in Bloom Once More（27）,(28)
Conviction: imitation of nature
Principle: printing and dyeing color, fabric texture
Method: 4 sides in continuation, zonal tying-in 21.6 cm, longitudinal back tying 32.1 cm
Operation: jacquard weaving, embroidery, sanding, washing, spraying dye
Trend: 2014/2015 fashion trend of Shanghai
Application: clothing fabrics, decorative cloth

Creative Fabrics Design 1

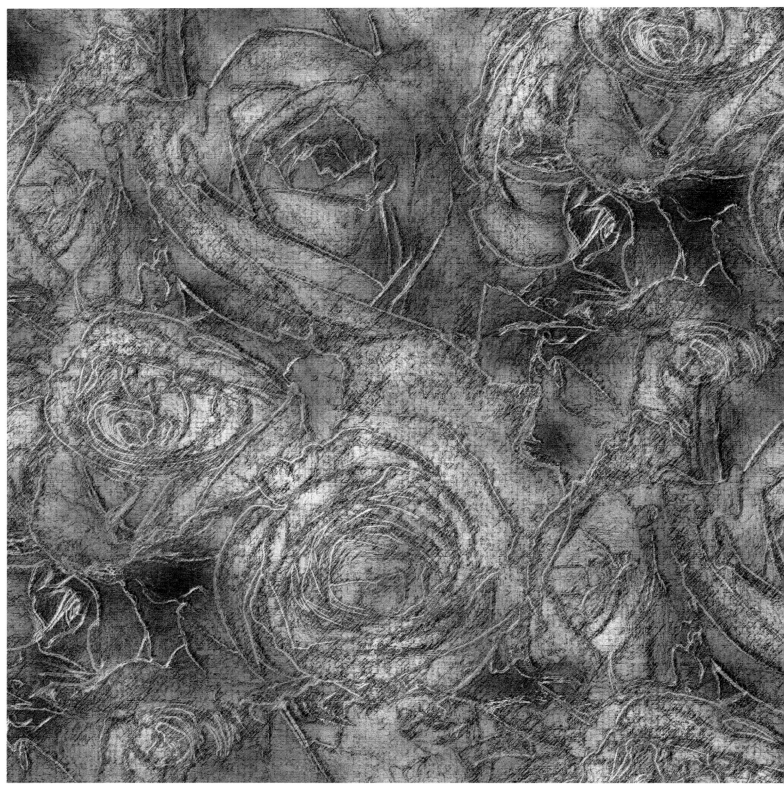

名：月季再逐艳（29）、（30）
道：道法自然
理：印染色蕴、织物组织层次
法：四方连缀、纬向接回21.6厘米、经向接回32.1厘米
术：梭织提花、皮革、刺绣、喷染、石磨、水洗
势：2014/2015海派流行趋势
器：服装面料、装饰布

Name: Rose in Bloom Once More (29), (30)
Conviction: imitation of nature
Principle: printing and dyeing color, fabric texture
Method: 4 sides in continuation, zonal tying-in 21.6 cm, longitudinal back tying 32.1 cm
Operation: jacquard weaving, embroidery, leathering, grinding, water washing, spraying dye
Trend: 2014/2015 fashion trend of Shanghai
Application: clothing fabrics, decorative cloth

Creative Fabrics Design 1

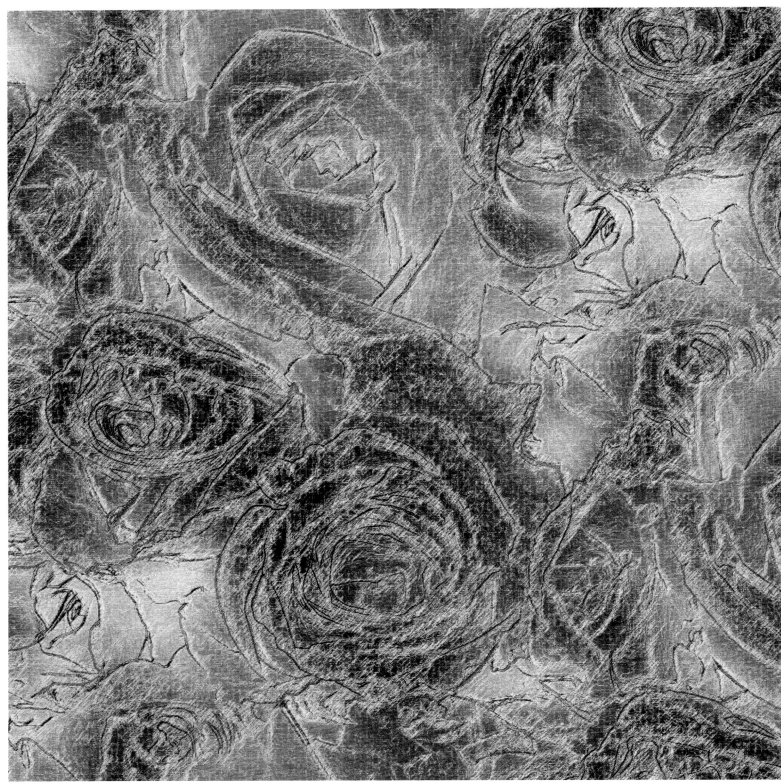

名：月季再逐艳（31）、（32）
道：道法自然
理：印染色蕴、织物组织层次
法：四方连缀、纬向接回21.6厘米、经向接回32.1厘米
术：皮革、梭织提花、刺绣、喷染、石磨、水洗
势：2014/2015海派流行趋势
器：服装面料、装饰布
Name: Rose in Bloom Once More（31）,（32）
Conviction: imitation of nature
Principle: printing and dyeing color, fabric texture
Method: 4 sides in continuation, zonal tying-in 21.6 cm, longitudinal back tying 32.1 cm
Operation: leathering, jacquard weaving, embroidery, spray-painting, grinding, water washing
Trend: 2014/2015 fashion trend of Shanghai
Application: clothing fabrics, decorative cloth

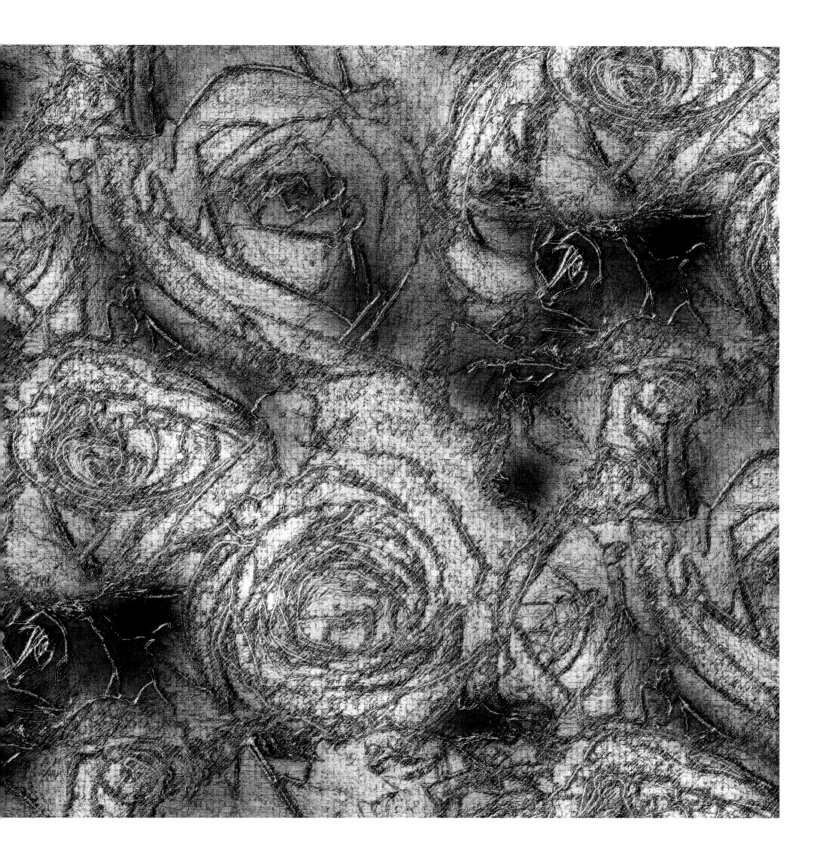

Creative Fabrics Design 1

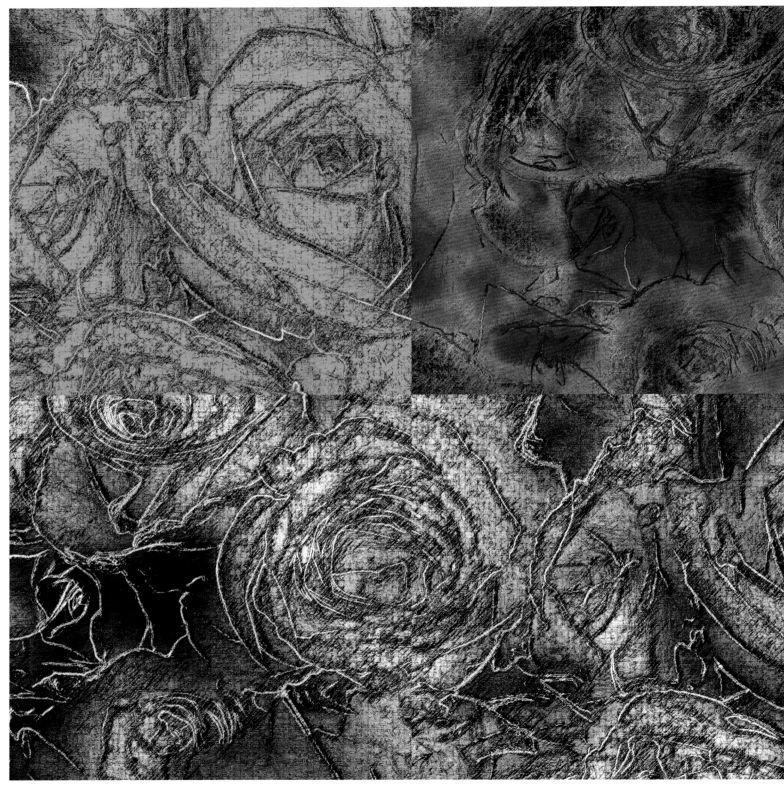

名：月季再逐艳 (33)、(34)、(35)、(36)、(37)
道：道法自然
理：印染色蕴、织物组织层次
法：四方连缀、纬向接回21.6厘米、经向接回32.1厘米
术：梭织提花、刺绣、喷染、石磨、水洗
势：2014/2015海派流行趋势
器：服装面料、装饰布

Name: Rose in Bloom Once More (33), (34), (35), (36), (37)
Conviction: imitation of nature
Principle: printing and dyeing color, fabric texture
Method: 4 sides in continuation, zonal tying-in 21.6 cm, longitudinal back tying 32.1 cm
Operation: jacquard weaving, embroidery, spray-painting, grinding, water washing
Trend: 2014/2015 fashion trend of Shanghai
Application: clothing fabrics, decorative cloth

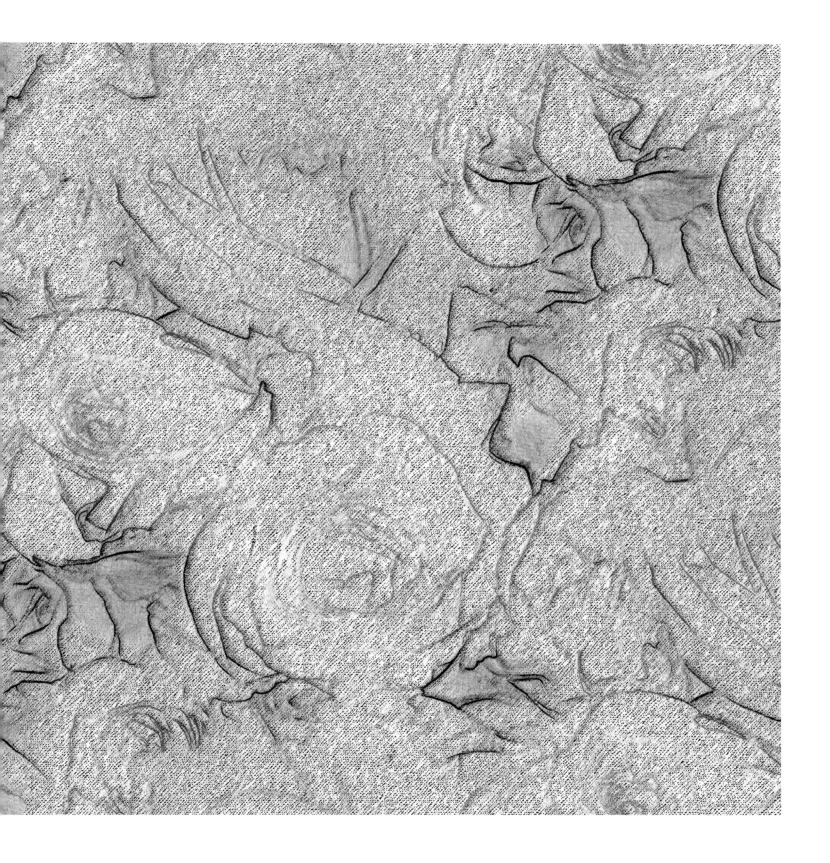

Creative Fabrics Design 1

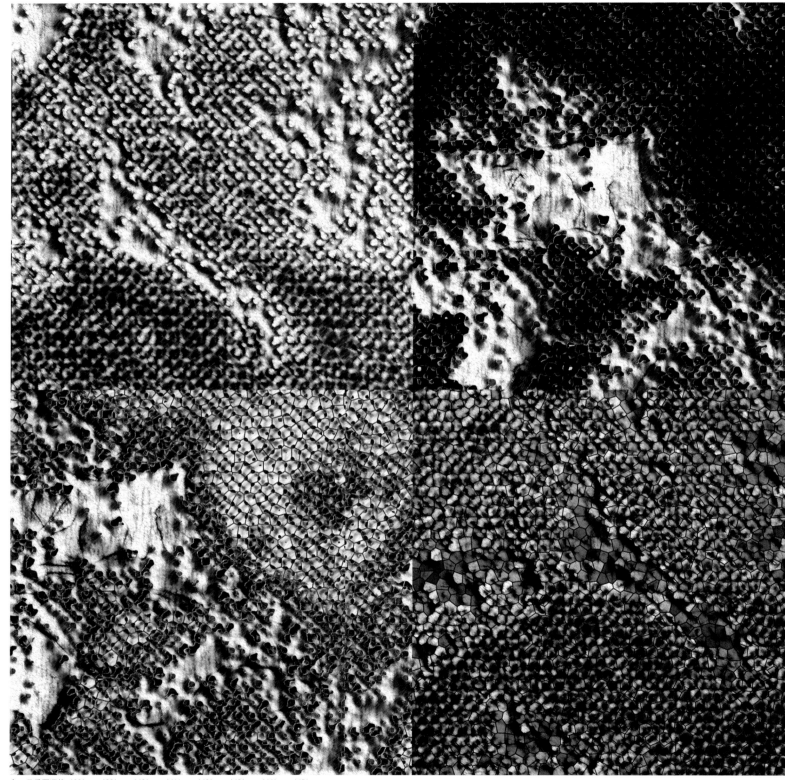

名：月季再逐艳 (38)、(39)、(40)、(41)、(42)、(43)、(44)、(45)
道：道法自然
理：印染色蕴、织物组织层次
法：四方连缀、纬向接回21.6厘米、经向接回32.1厘米
术：梭织提花、起皱纱线、加穗纱线、拉绒、粘合
势：2014/2015海派流行趋势
器：服装面料、装饰布

Name: Rose in Bloom Once More (38), (39), (40), (41), (42), (43), (44), (45)
Conviction: imitation of nature
Principle: printing and dyeing color, fabric texture
Method: 4 sides in continuation, zonal tying-in 21.6 cm, longitudinal back tying 32.1 cm
Operation: jacquard weaving, wrinkling yarn, fringing yarn, napping, agglutinating
Trend: 2014/2015 fashion trend of Shanghai
Application: clothing fabrics, decorative cloth

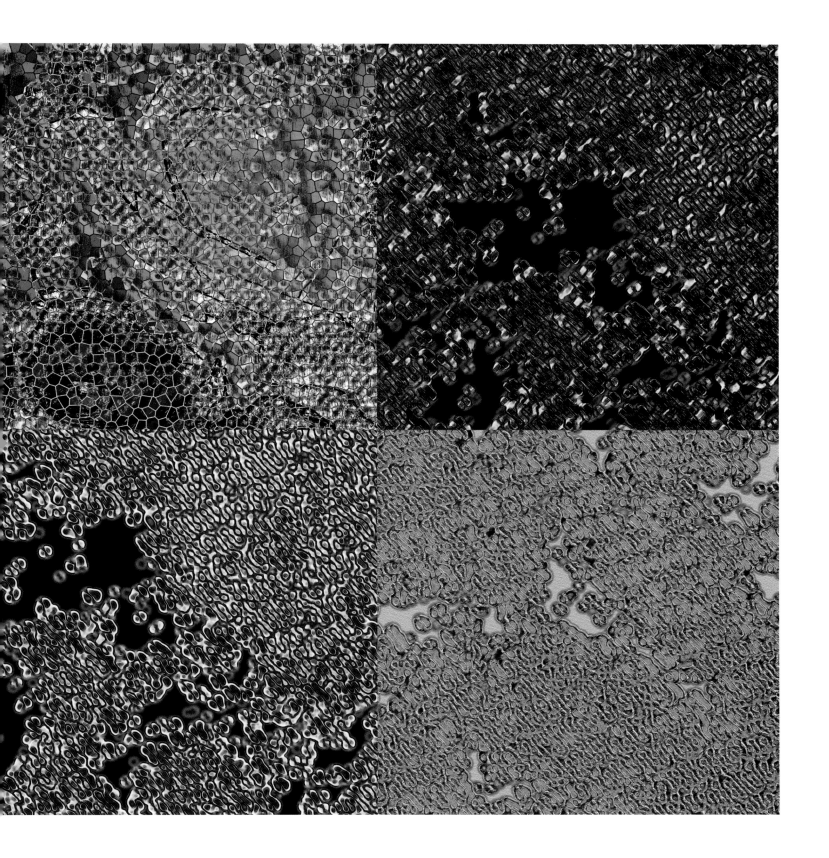

Creative Fabrics Design 1

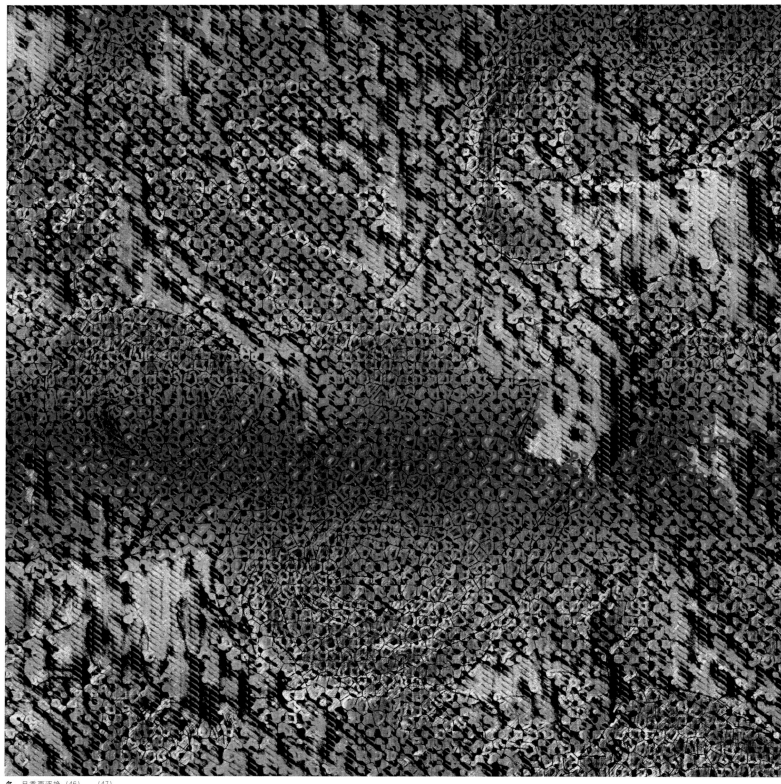

名：月季再逐艳（46）、（47）
道：道法自然
理：印染色蕴、织物组织层次
法：四方连缀、纬向接回21.6厘米、经向接回32.1厘米
术：梭织提花、起皱纱线、加穗纱线、拉绒、粘合
势：2014/2015海派流行趋势
器：服装面料、装饰布

Name: Rose in Bloom Once More (46), (47)
Conviction: imitation of nature
Principle: printing and dyeing color, fabric texture
Method: 4 sides in continuation, zonal tying-in 21.6 cm, longitudinal back tying 32.1 cm
Operation: jacquard weaving, wrinkling yarn, fringing yarn, napping, agglutinating
Trend: 2014/2015 fashion trend of shanghai
Application: clothing fabrics, decorative cloth

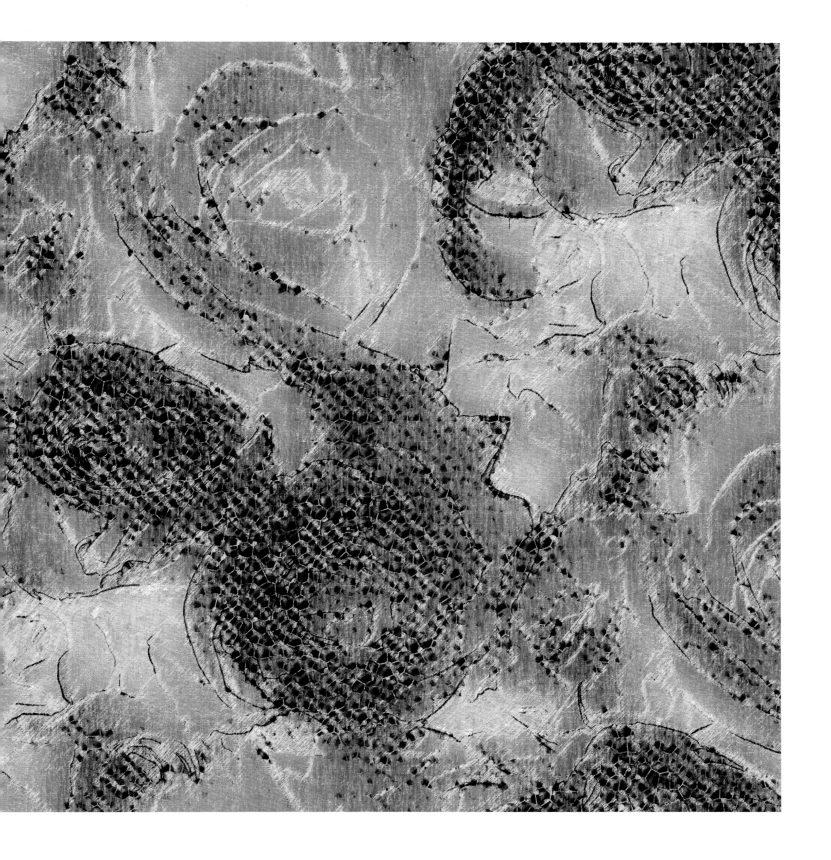

Creative Fabrics Design 1

名：玫菲瑰舞（8）、（9）、（10）、（11）
道：东西方民俗混搭
理：银蚀刻、瓷刻肌理
法：四方连缀，纬向接回35厘米，经向接回64.2厘米
术：梭织提花、刻花、镂空、涂层
势：2014/2015国际流行趋之势艺术家表现形态
器：服装面料、装饰布

Name: Rose Rose Dance (8), (9), (10), (11)
Conviction: culture mixture of East and West
Principle: silver etching, porcelain carved texture
Method: 4 sides in continuation, zonal tying-in 35 cm, longitudinal tying-in 64.2 cm
Operation: jacquard weaving, carving, hollowing, coating
Trend: 2014/2015 international fashion trend of artists expression
Application: clothing fabrics, decorative cloth

Creative Fabrics Design 1

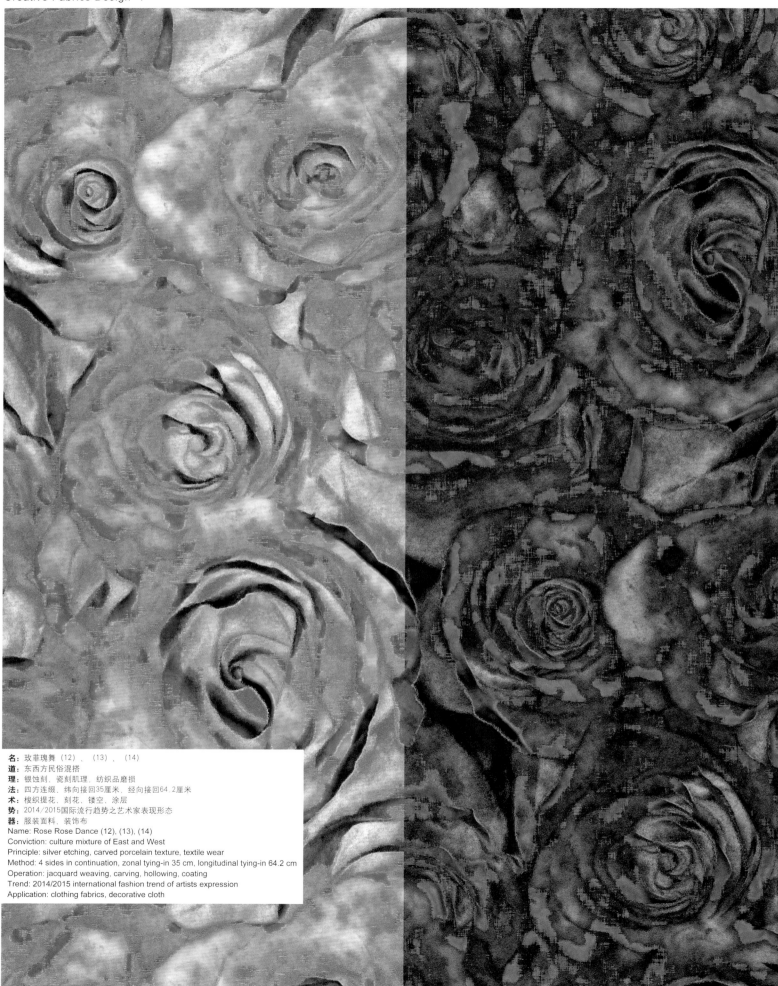

名：玫菲瑰舞（12）、（13）、（14）
道：东西方民俗混搭
理：银蚀刻、瓷刻肌理、纺织品磨损
法：四方连缀、纬向接回35厘米、经向接回64.2厘米
术：梭织提花、刻花、镂空、涂层
势：2014/2015国际流行趋势之艺术家表现形态
器：服装面料、装饰布
Name: Rose Rose Dance (12), (13), (14)
Conviction: culture mixture of East and West
Principle: silver etching, carved porcelain texture, textile wear
Method: 4 sides in continuation, zonal tying-in 35 cm, longitudinal tying-in 64.2 cm
Operation: jacquard weaving, carving, hollowing, coating
Trend: 2014/2015 international fashion trend of artists expression
Application: clothing fabrics, decorative cloth

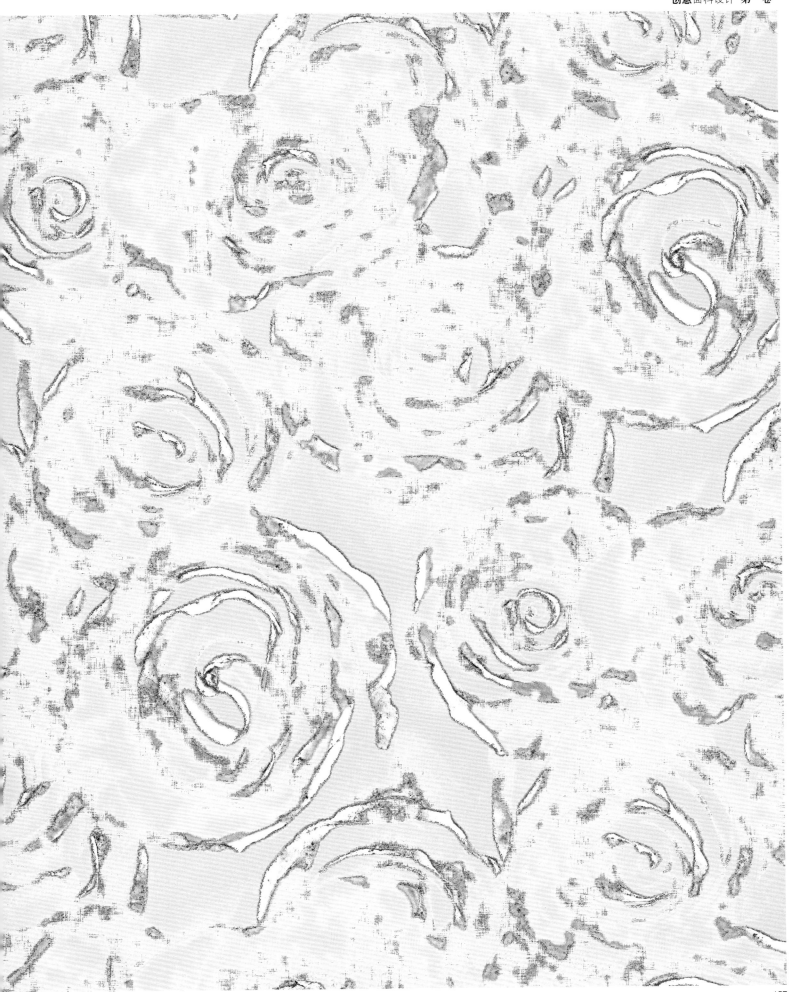

Creative Fabrics Design 1

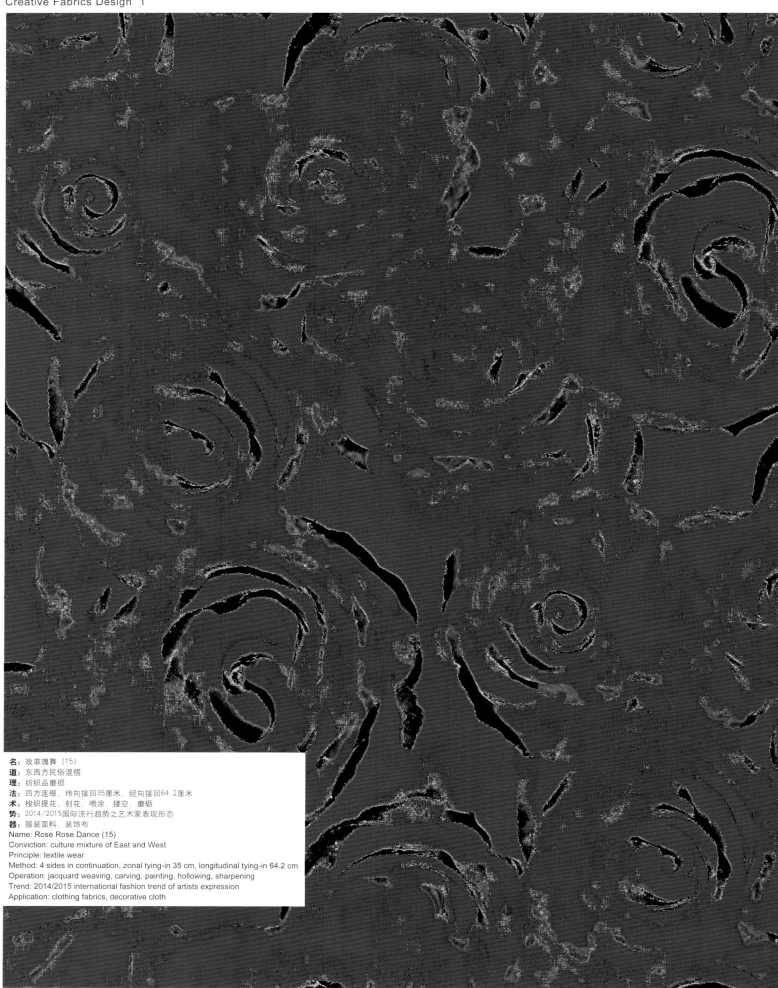

名：玫菲瑰舞（15）
道：东西方民俗混搭
理：纺织品磨损
法：四方连缀，纬向接回35厘米，经向接回64.2厘米
术：梭织提花、刻花、喷涂、镂空、磨砺
势：2014/2015国际流行趋势之艺术家表现形态
器：服装面料，装饰布
Name: Rose Rose Dance (15)
Conviction: culture mixture of East and West
Principle: textile wear
Method: 4 sides in continuation, zonal tying-in 35 cm, longitudinal tying-in 64.2 cm
Operation: jacquard weaving, carving, painting, hollowing, sharpening
Trend: 2014/2015 international fashion trend of artists expression
Application: clothing fabrics, decorative cloth

时间：2004年7月18日
地点：四川稻城
对象：矿石
灵感：幻影
Time: July 18, 2004
Location: Daocheng, Sichuan
Object: ore
Inspiration: fragmented and mottled

Creative Fabrics Design 1

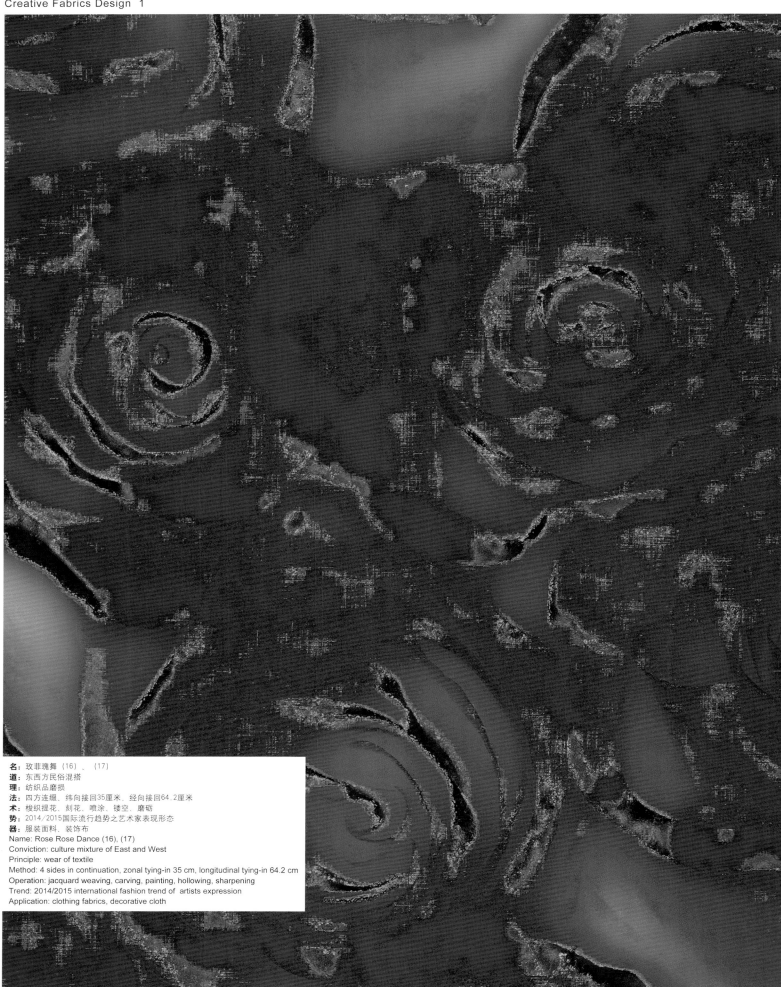

名：玫菲瑰舞（16）、（17）
道：东西方民俗混搭
理：纺织品磨损
法：四方连缀、纬向接回35厘米，经向接回64.2厘米
术：梭织提花、刻花、喷涂、镂空、磨砺
势：2014/2015国际流行趋势之艺术家表现形态
器：服装面料、装饰布
Name: Rose Rose Dance (16), (17)
Conviction: culture mixture of East and West
Principle: wear of textile
Method: 4 sides in continuation, zonal tying-in 35 cm, longitudinal tying-in 64.2 cm
Operation: jacquard weaving, carving, painting, hollowing, sharpening
Trend: 2014/2015 international fashion trend of artists expression
Application: clothing fabrics, decorative cloth

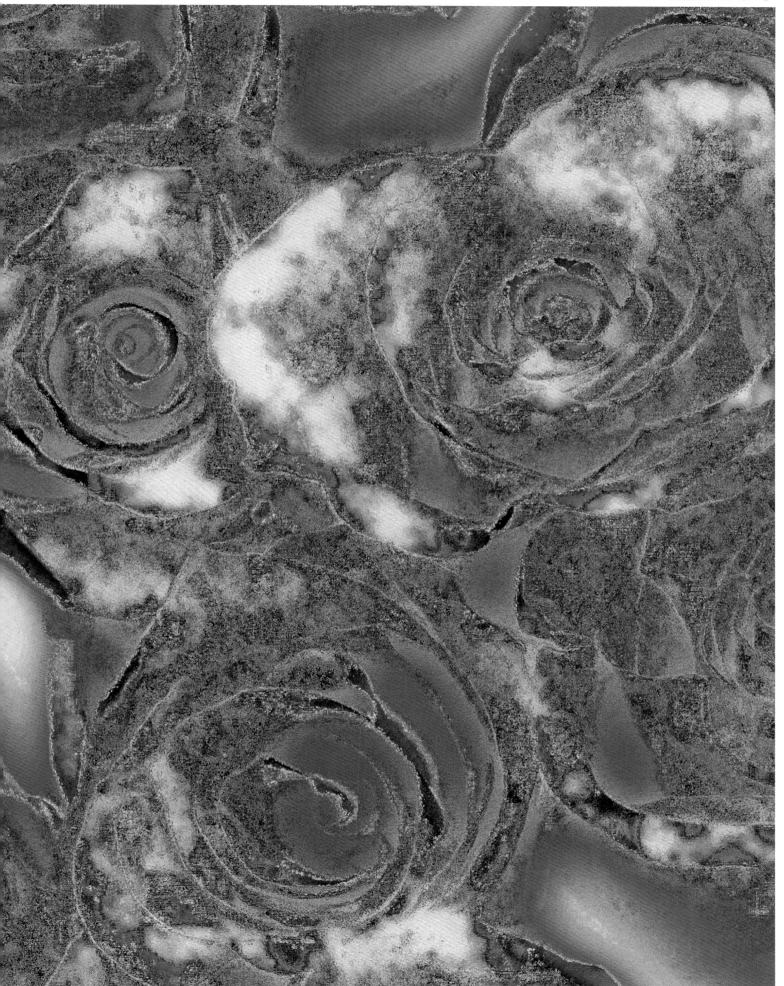

Creative Fabrics Design 1

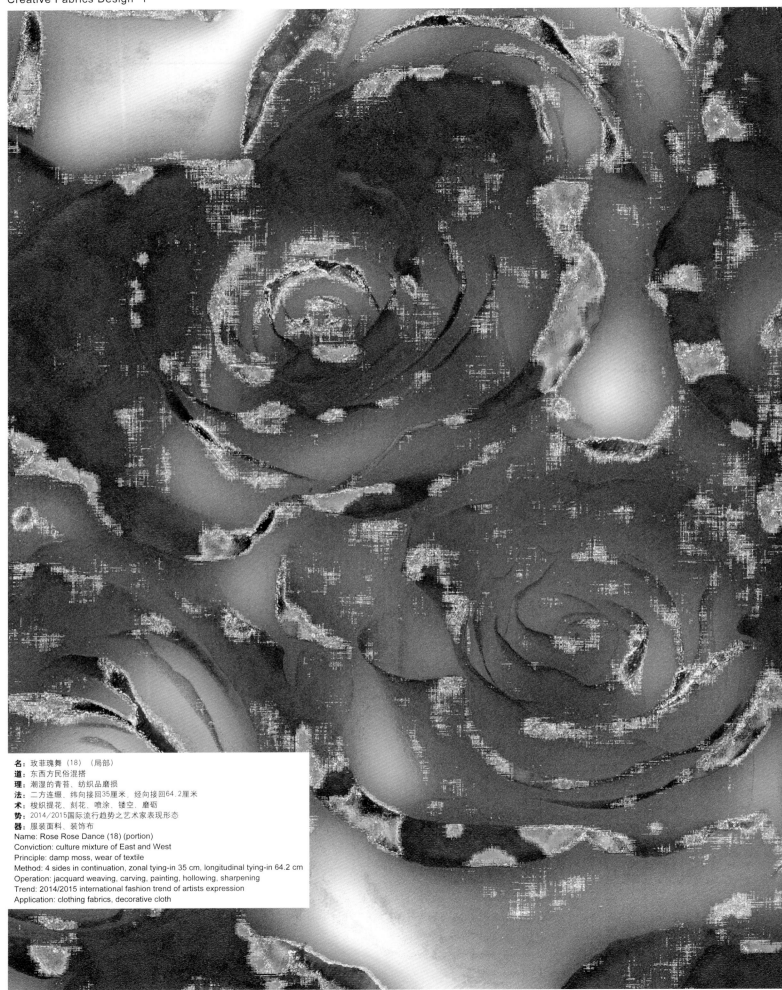

名：玫菲瑰舞（18）（局部）
道：东西方民俗混搭
理：潮湿的青苔、纺织品磨损
法：二方连缀、纬向接回35厘米，经向接回64.2厘米
术：梭织提花、刻花、喷涂、镂空、磨砺
势：2014/2015国际流行趋势之艺术家表现形态
器：服装面料、装饰布

Name: Rose Rose Dance (18) (portion)
Conviction: culture mixture of East and West
Principle: damp moss, wear of textile
Method: 4 sides in continuation, zonal tying-in 35 cm, longitudinal tying-in 64.2 cm
Operation: jacquard weaving, carving, painting, hollowing, sharpening
Trend: 2014/2015 international fashion trend of artists expression
Application: clothing fabrics, decorative cloth

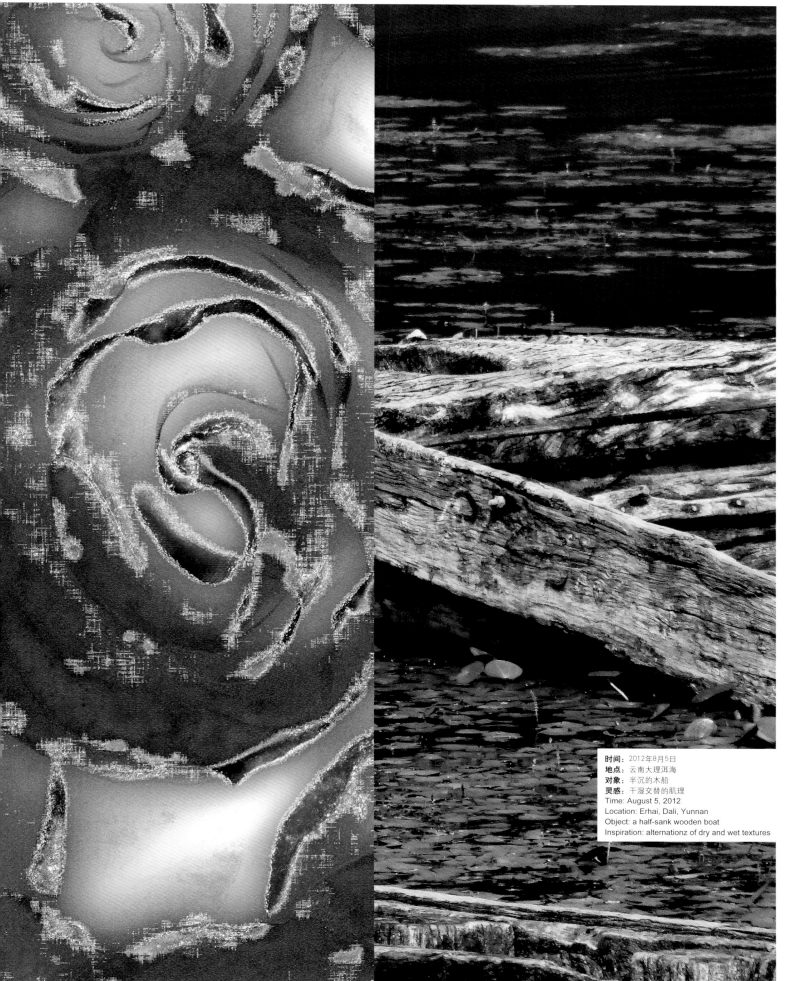

时间：2012年8月5日
地点：云南大理洱海
对象：半沉的木船
灵感：干湿交替的肌理
Time: August 5, 2012
Location: Erhai, Dali, Yunnan
Object: a half-sank wooden boat
Inspiration: alternationz of dry and wet textures

Creative Fabrics Design 1

名：芳菲升腾（23）、（24）
道：道法自然
理：印染色蕴、织物组织层次
法：四方连缀、纬向接回35厘米、经向接回64.2厘米
术：梭织提花，烂花
势：2014/2015国际流行趋势之自然生物形态
器：服装面料、装饰布

Name: Rose in Love (23), (24)
Conviction: imitation of nature
Principle: printing and dyeing color, fabric texture
Method: 4 sides in continuation, zonal tying-in 35 cm, longitudinal tying-in 64.2 cm
Operation: jacquard weaving, burnt-discharging
Trend: 2014/2015 international fashion trend of bio feature
Application: clothing fabrics, decorative cloth

时间：2010年10月22日
地点：尼泊尔本迪布尔
对象：仙人掌
灵感：争奇斗艳般竞逐生长
Time: October 22, 2010
Location: The Dibble, Nepal
Object: cactus
Inspiration: cactas competing for growth

Creative Fabrics Design 1

名： 君临（27）、（31）
道： 东西方传说混搭
理： 织物涂层光晕
法： 二方连缀、纬向接回70厘米，经向定位85厘米
术： 梭织提花、刻花、喷涂、镂空、磨砺
势： 2014/2015秋冬海派流行趋势
器： 服装面料、装饰布

Name: Kingly Landing (27), (31)
Conviction: culture mixture of East and West
Principle: fabric coating shading
Method: 2 sides in continuation, zonal tying-in 70 cm, longitudinal tying-in, 85cm
Operation: jacquard weaving, carving, painting, hollowing, sharpening
Trend: 2014/2015 autumn and winter fashion trend of Shanghai
Application: clothing fabrics, decorative cloth

Creative Fabrics Design 1

名：元旦巴德岗（7）
道：宗教艺术
理：艳阳下的石雕
法：四方连缀，纬向接回35厘米，经向接回64.2厘米
术：梭织提花，特印，喷涂
势：2014/2015国际流行趋势之民族文化形态
器：服装面料，装饰布

Name: Bhaktapur on New Year's Day (7)
Conviction: religious art
Principle: stone carving under the sun
Method: 4 sides in continuation, zonal tying-in 35 cm, longitudinal tying-in 64.2 cm
Operation: jacquard weaving, printing, spraying
Trend: 2014/2015 international fashion trend of national culture
Application: clothing fabrics, decorative cloth

时间：2010年10月26日
地点：尼泊尔巴德岗
对象：石塔
灵感：宗教艺术
Time: October 26, 2010
Location: Bhaktapur, Nepal
Object: stone tower
Inspiration: religious art

Creative Fabrics Design 1

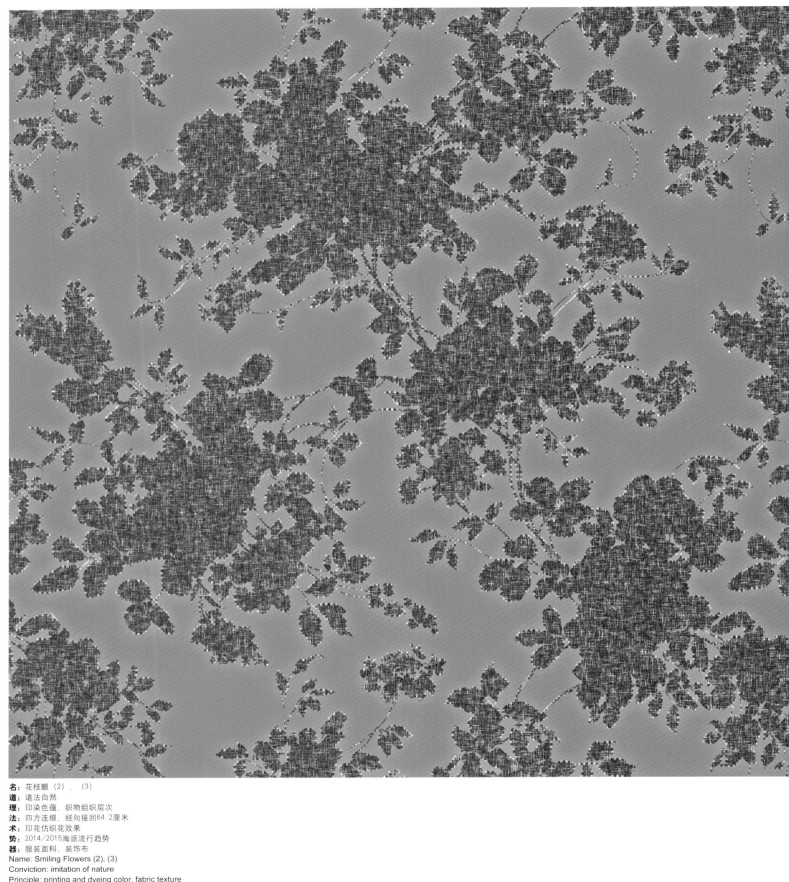

名：花枝颤（2）、（3）
道：道法自然
理：印染色蕴、织物组织层次
法：四方连缀，经向接回64.2厘米
术：印花仿织花效果
势：2014/2015海派流行趋势
器：服装面料、装饰布

Name: Smiling Flowers (2), (3)
Conviction: imitation of nature
Principle: printing and dyeing color, fabric texture
Method: 4 sides in continuation, zonal tying-in 21.6 cm, longitudinal back tying 32.1 cm
Operation: printing, in imitation of the effect of woven pattern
Trend: 2014/2015 fashion trend of Shanghai
Application: clothing fabrics, decorative cloth

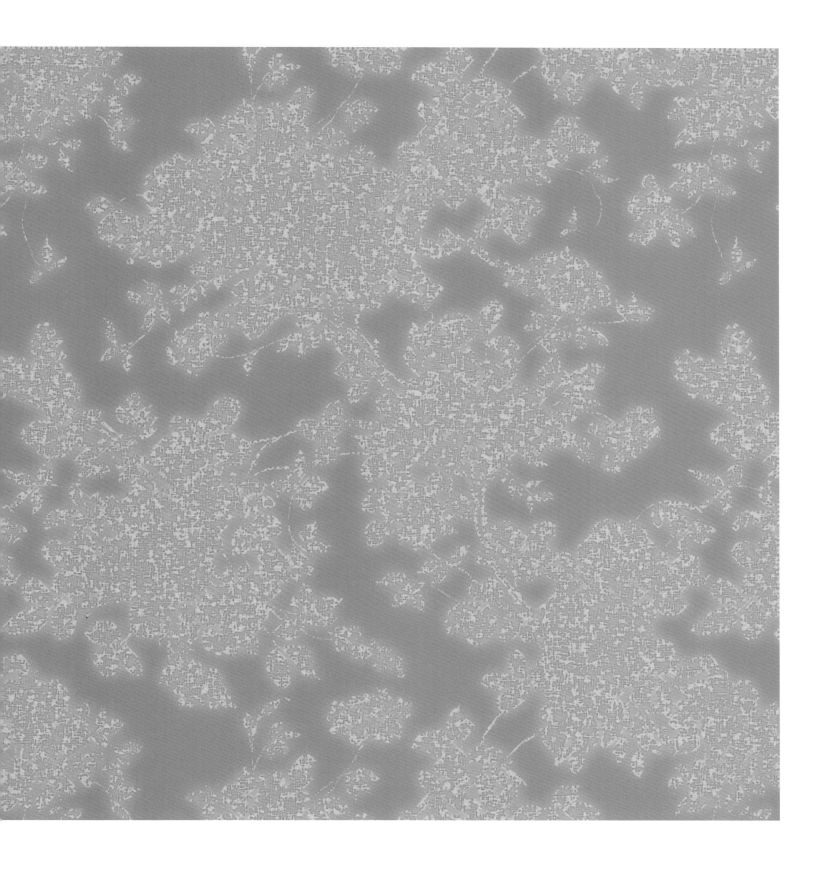

Creative Fabrics Design 1

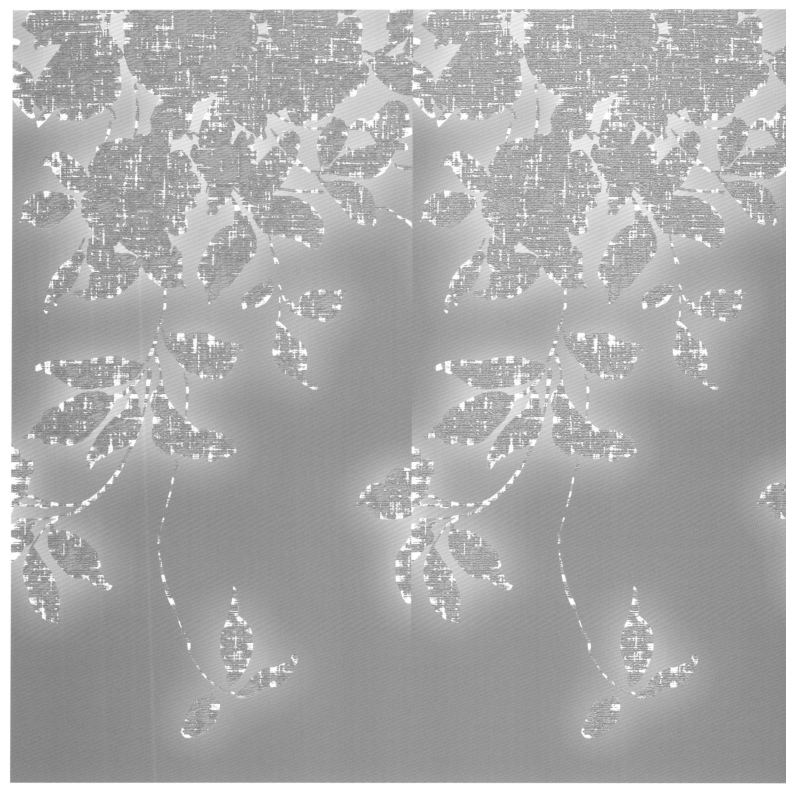

名：花枝颤（4）、（5）、（6）、（7）
道：道法自然
理：印染色蕴、织物组织层次
法：四方连缀、纬向接回21.6厘米、经向接回32.1厘米
术：印花仿织花效果
势：2014/2015海派流行趋势
器：服装面料、装饰布
Name: Smiling Flowers (4), (5), (6), (7)
Conviction: imitation of nature
Principle: printing and dyeing color, fabric texture
Method: 4 sides in continuation, longitudinal tying back 64.2cm
Operation: printing, in imitation of the effect of woven pattern
Trend: 2014/2015 fashion trend of Shanghai
Application: clothing fabrics, decorative cloth

Creative Fabrics Design 1

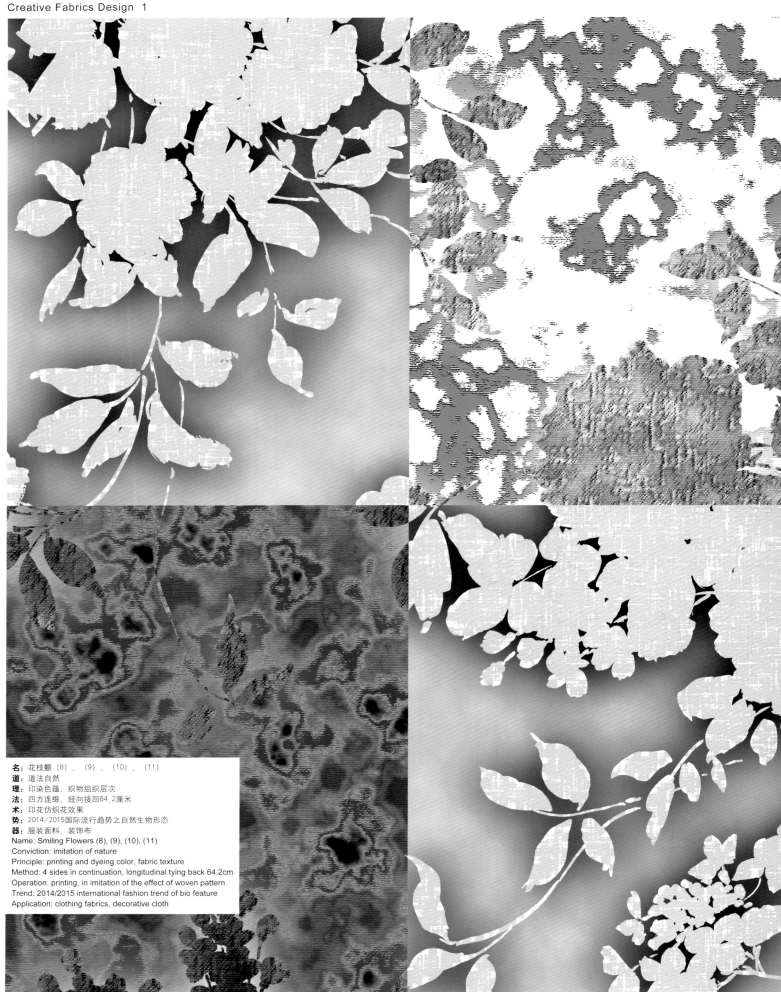

名：花枝颤（8）、（9）、（10）、（11）
道：道法自然
理：印染色蕴、织物组织层次
法：四方连缀、经向接回64.2厘米
术：印花仿织花效果
势：2014/2015国际流行趋势之自然生物形态
器：服装面料、装饰布
Name: Smiling Flowers (8), (9), (10), (11)
Conviction: imitation of nature
Principle: printing and dyeing color, fabric texture
Method: 4 sides in continuation, longitudinal tying back 64.2cm
Operation: printing, in imitation of the effect of woven pattern
Trend: 2014/2015 international fashion trend of bio feature
Application: clothing fabrics, decorative cloth

Creative Fabrics Design 1

名：花枝颤（12）、（13）、（14）、（16）、（17）、（18）、（19）
道：道法自然
理：印染色蕴、织物组织层次
法：四方连缀、经向接回64.2厘米
术：印花仿织花效果
势：2014/2015国际流行趋势之自然生物形
器：服装面料、装饰布
Name: Smiling Flowers (12), (13), (14), (16), (17), (18), (19)
Conviction: imitation of nature
Principle: printing and dyeing color, fabric texture
Method: 4 sides in continuation, longitudinal tying back 64.2cm
Operation: printing, in imitation of the effect of woven pattern
Trend: 2014/2015 international fashion trend of bio feature
Application: clothing fabrics, decorative cloth

Creative Fabrics Design 1

名：花枝颤（15）
道：道法自然
理：印染色蕴、织物组织层次
法：四方连缀、经向接回64.2厘米
术：印花仿织花效果
势：2014/2015国际流行趋势之自然生物形态
器：服装面料、装饰布

Name: Smiling Flowers (15)
Conviction: imitation of nature
Principle: printing and dyeing color, fabric texture
Method: 4 sides in continuation, longitudinal tying back 64.2cm
Operation: printing, in imitation of the effect of woven pattern
Trend: 2014/2015 international fashion trend of bio feature
Application: clothing fabrics, decorative cloth

Creative Fabrics Design 1

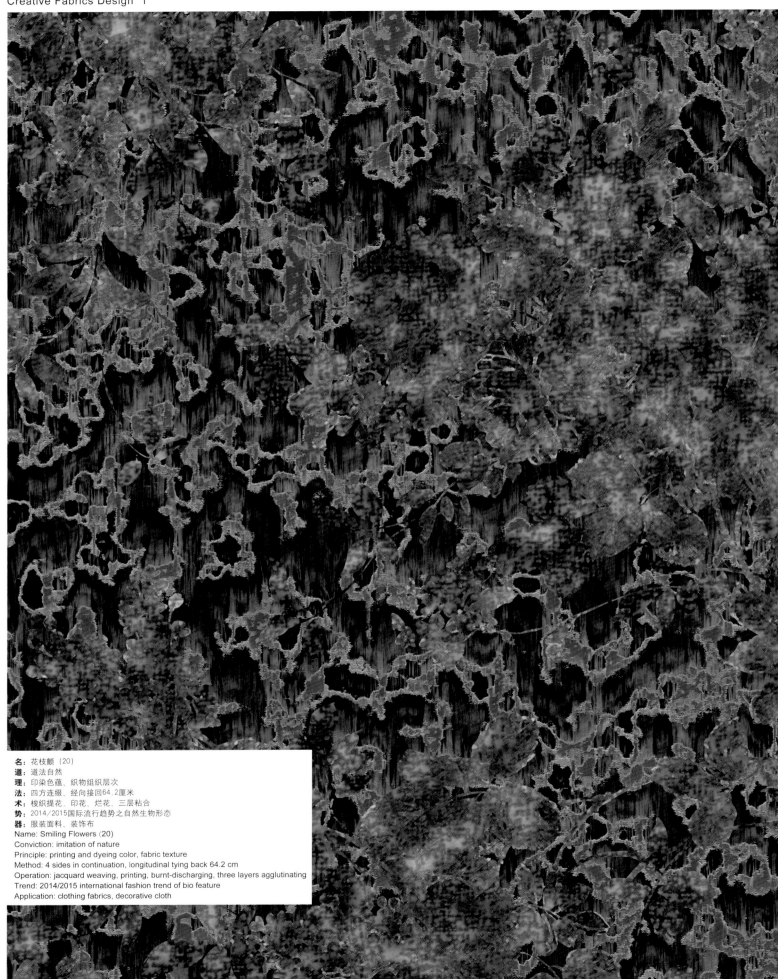

名：花枝颤（20）
道：道法自然
理：印染色蕴、织物组织层次
法：四方连缀、经向接回64.2厘米
术：梭织提花、印花、烂花、三层粘合
势：2014/2015国际流行趋势之自然生物形态
器：服装面料、装饰布

Name: Smiling Flowers (20)
Conviction: imitation of nature
Principle: printing and dyeing color, fabric texture
Method: 4 sides in continuation, longitudinal tying back 64.2 cm
Operation: jacquard weaving, printing, burnt-discharging, three layers agglutinating
Trend: 2014/2015 international fashion trend of bio feature
Application: clothing fabrics, decorative cloth

Creative Fabrics Design 1

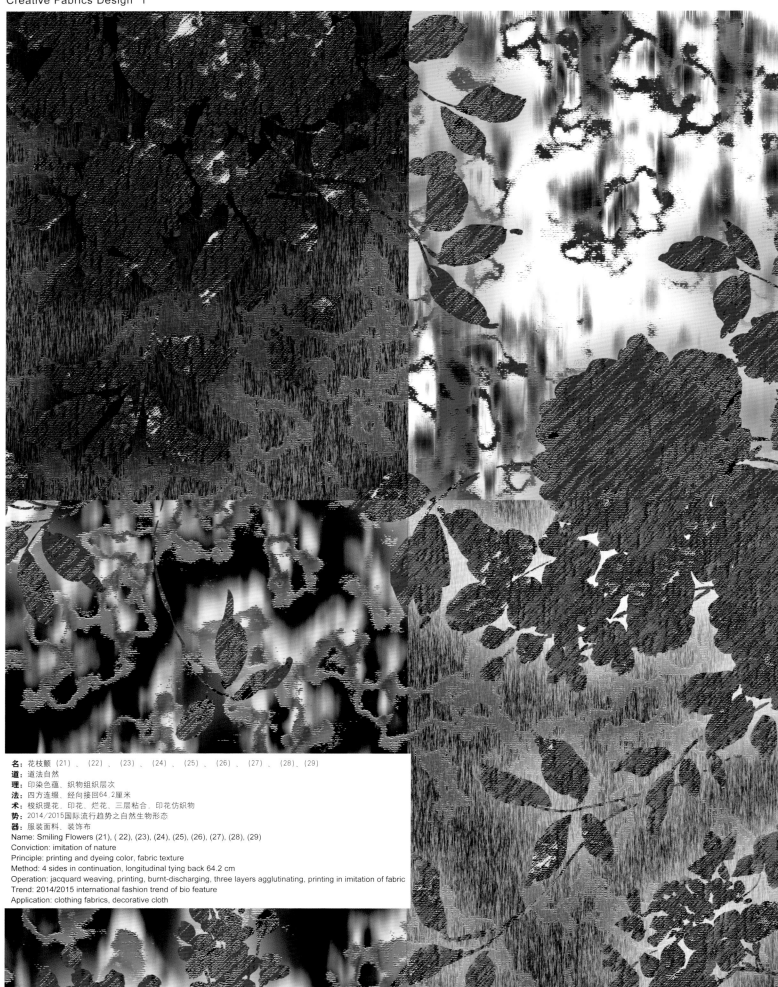

名：花枝颤（21）、（22）、（23）、（24）、（25）、（26）、（27）、（28）、（29）
道：道法自然
理：印染色蕴、织物组织层次
法：四方连缀、经向接回64.2厘米
术：梭织提花、印花、烂花、三层粘合、印花仿织物
势：2014/2015国际流行趋势之自然生物形态
器：服装面料、装饰布
Name: Smiling Flowers (21), (22), (23), (24), (25), (26), (27), (28), (29)
Conviction: imitation of nature
Principle: printing and dyeing color, fabric texture
Method: 4 sides in continuation, longitudinal tying back 64.2 cm
Operation: jacquard weaving, printing, burnt-discharging, three layers agglutinating, printing in imitation of fabric
Trend: 2014/2015 international fashion trend of bio feature
Application: clothing fabrics, decorative cloth

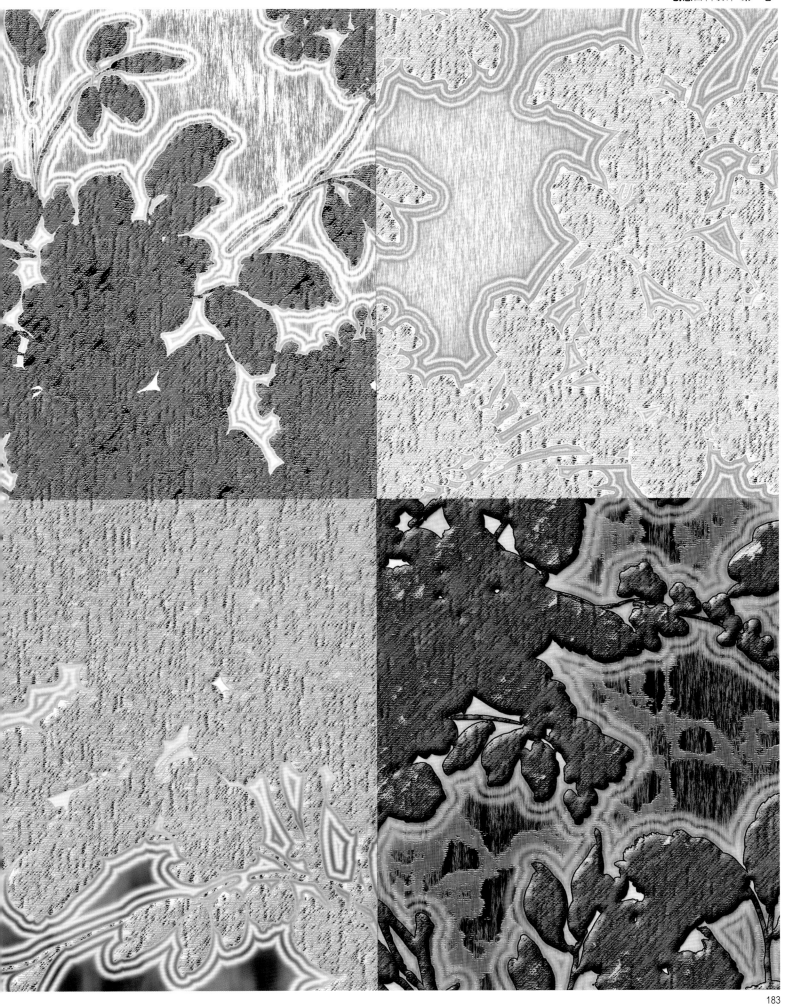

名：花枝颤（30）
道：道法自然
理：印染色蕴、织物组织层次
法：四方连缀、经向接回64.2厘米
术：梭织提花、印花、烂花、二层粘合、印花仿织物
势：2014/2015国际流行趋势之自然生物形态
器：服装面料、装饰布

Name: Smiling Flowers (30)
Conviction: imitation of nature
Principle: printing and dyeing color, fabric texture
Method: 4 sides in continuation, longitudinal tying back 64.2 cm
Operation: jacquard weaving, printing, burnt-discharging, two layers agglutinating, printing in imitation of fabric
Trend: 2014/2015 international fashion trend of bio feature
Application: clothing fabrics, decorative cloth

时间：2010年8月4日
地点：四川西昌邛海
对象：枝梢
灵感：折枝花形式
Time: August 4, 2010
Location: Qionghai Lake, Xichang, Sichuan
Object: branches
Inspiration: sprays of flowers

Creative Fabrics Design 1

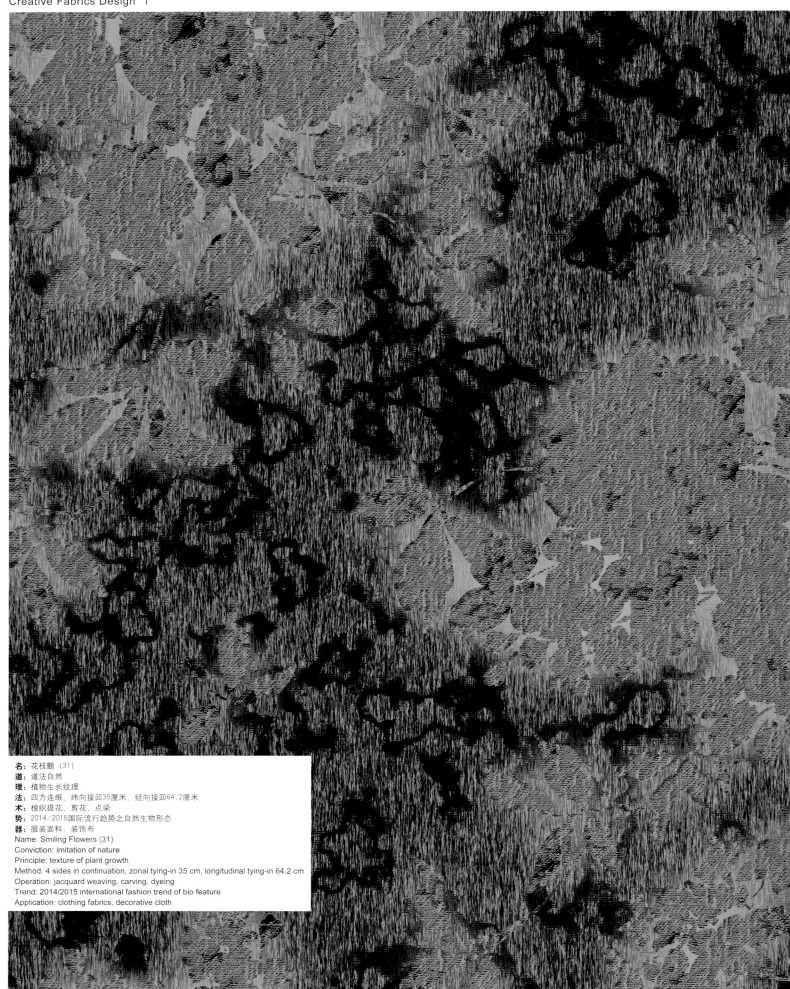

名：花枝颤（31）
道：道法自然
理：植物生长纹理
法：四方连缀、纬向接回35厘米、经向接回64.2厘米
术：梭织提花、剪花、点染
势：2014/2015国际流行趋势之自然生物形态
器：服装面料、装饰布
Name: Smiling Flowers (31)
Conviction: imitation of nature
Principle: texture of plant growth
Method: 4 sides in continuation, zonal tying-in 35 cm, longitudinal tying-in 64.2 cm
Operation: jacquard weaving, carving, dyeing
Trend: 2014/2015 international fashion trend of bio feature
Application: clothing fabrics, decorative cloth

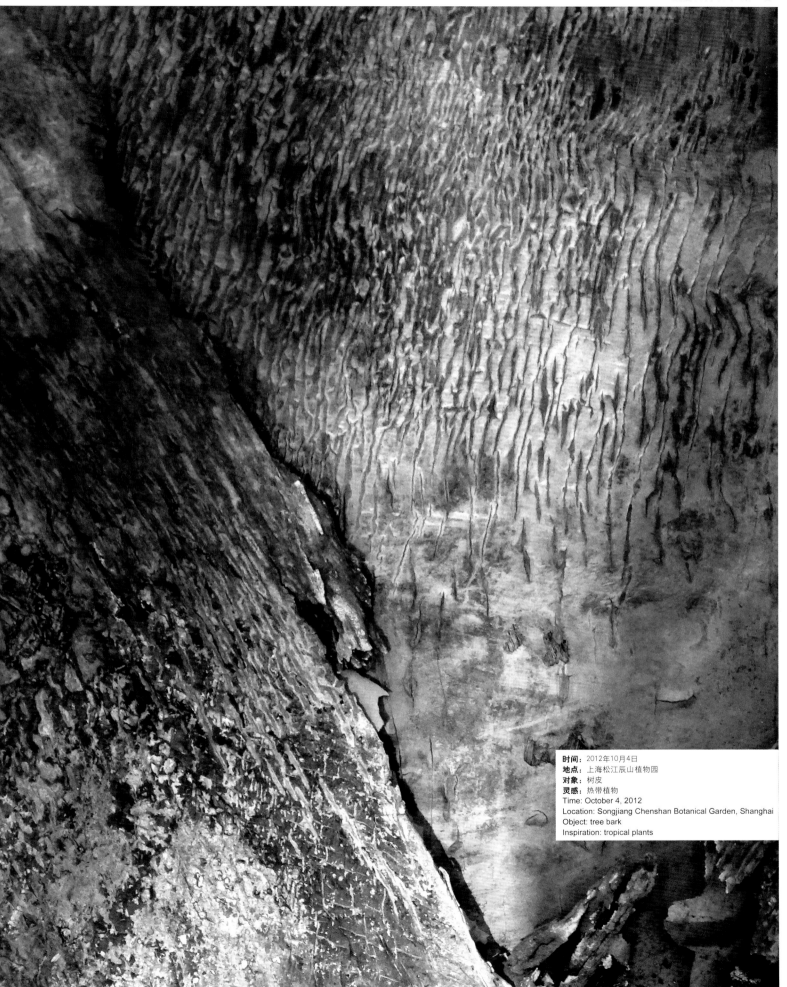
时间：2012年10月4日
地点：上海松江辰山植物园
对象：树皮
灵感：热带植物
Time: October 4, 2012
Location: Songjiang Chenshan Botanical Garden, Shanghai
Object: tree bark
Inspiration: tropical plants

名：华清池系列之瓷绘笔意（1）（局部与整体）
道：幻想
理：墨晕、织物组织层次
法：四方连续、纬向接回75厘米、经向接回160厘米
术：印花、植绒
势：2014/2015国际流行趋势之传统文化形态
器：服装面料、装饰布

Name: Huaqing Hot Spring Series: Ceramic Painting in Freehand Brushwork (1) (portion and whole)
Conviction: fantasy
Principle: ink shading, fabric texture
Method: 4 sides in continuation, zonal tying-in 75 cm, longitudinal tying-in 160 cm
Operation: printing, flocking
Trend: 2014/2015 international trend of traditional culture
Application: clothing fabrics, decorative cloth

Creative Fabrics Design 1

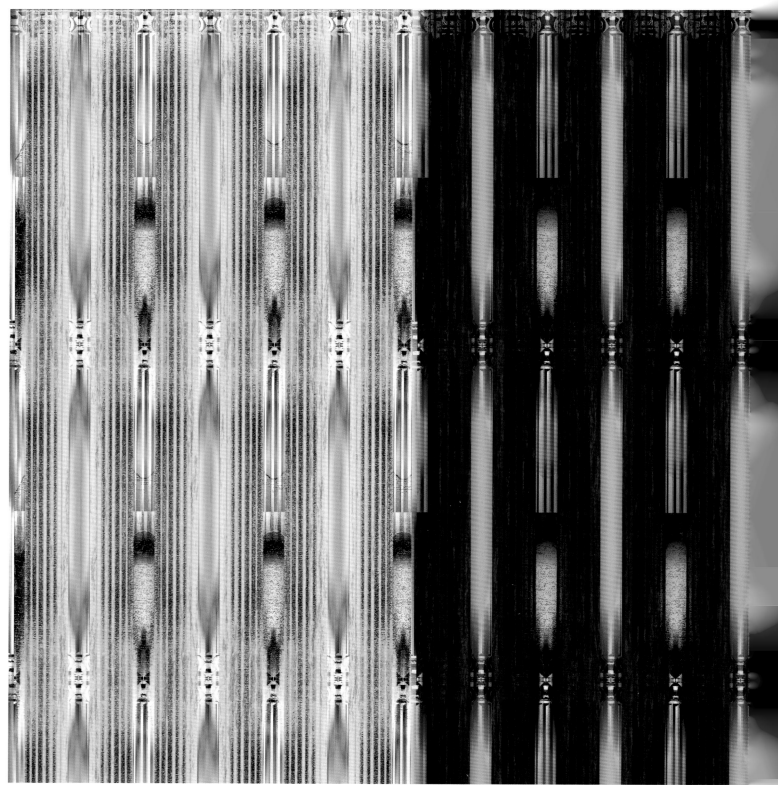

名：站立者的历史（11）、（16）、（26）
道：西风东进
理：风化、表面颗粒质
法：四方连缀、纬向接回10.8厘米、经向85厘米定位
术：梭织提花、烂花、烧毛、植绒
势：2014/2015秋冬流行趋势之外滩印象
器：服装面料、装饰布

Name: In the Winds of History (11), (16), (26)
Conviction: western impact on the east
Principle: weathering, surface particles
Method: 4 sides in continuation, zonal tying-in 10.8 cm, longitudinal positioning 85 cm
Operation: jacquard weaving, burnt-discharge, singeing, flocking
Trend: 2014/2015 autumn and winter fashion trend of Shanghai — impression of the Bund
Application: clothing fabrics, decorative cloth

时间：2008年2月21日
地点：上海外滩外白渡桥
对象：桥头
灵感：百年来的骄傲和沧桑
Time: February 21,2008
Location: Waibaidu Bridge of the Bund,Shanghai
Object: the bridge
Inspiration: a century of pride and vicissitudes

Creative Fabrics Design 1

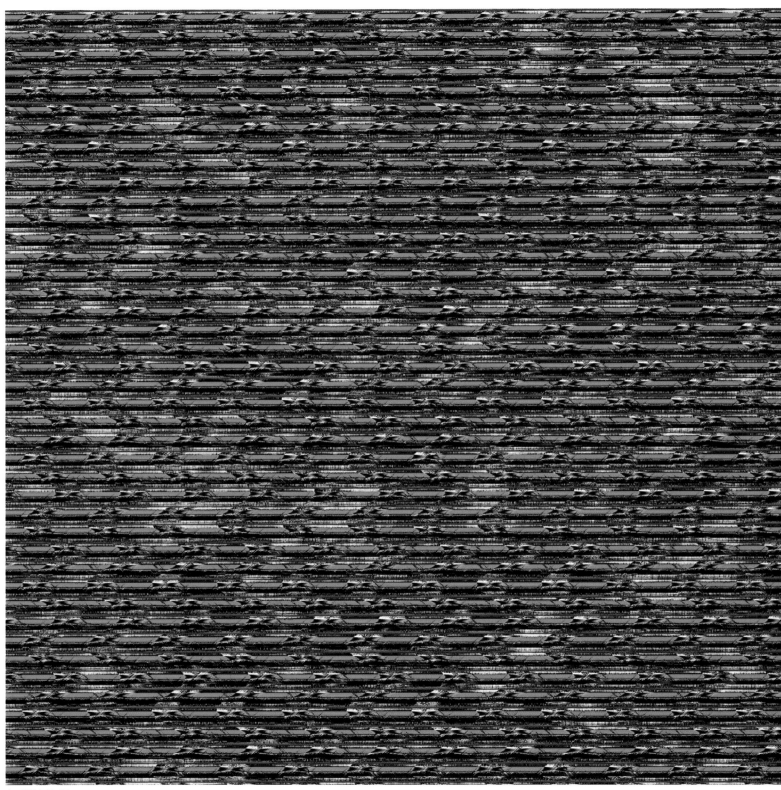

名：拨弦自诩（1）
道：天工开物
理：次序、表面吸潮质感
法：四方连缀，纬向接回10.8厘米，经向85厘米定位
术：梭织提花，针织提花
势：2014/2015秋冬国际流行趋势之民族文化形态
器：服装面料

Name: Self-Praising by Playing a Tune (1)
Conviction: heavenly creations
Principle: order texture, moisture absorption
Method: 4 sides in continuation, zonal tying-in 10.8 cm, longitudinal positioning 85 cm
Operation: jacquard weaving, jacquard knitting
Trend: 2014/2015 international fashion trend of national culture
Application: clothing fabrics

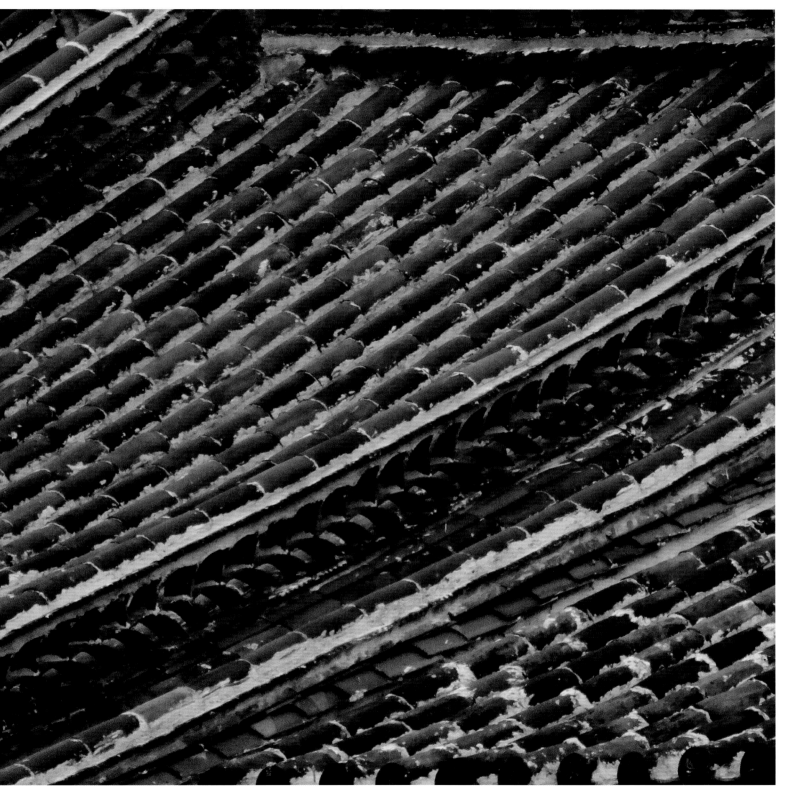

时间：2012年8月6日
地点：云南丽江黑龙潭
对象：屋顶瓦片
灵感：高原上的青瓦次序与质感
Device: clothing fabrics
Time: August 6, 2012
Location: Black Dragon Pool, Lijiang, Yunnan
Object: roof tiles
Inspiration: order and texture of grey tiles on the plateau

Creative Fabrics Design 1

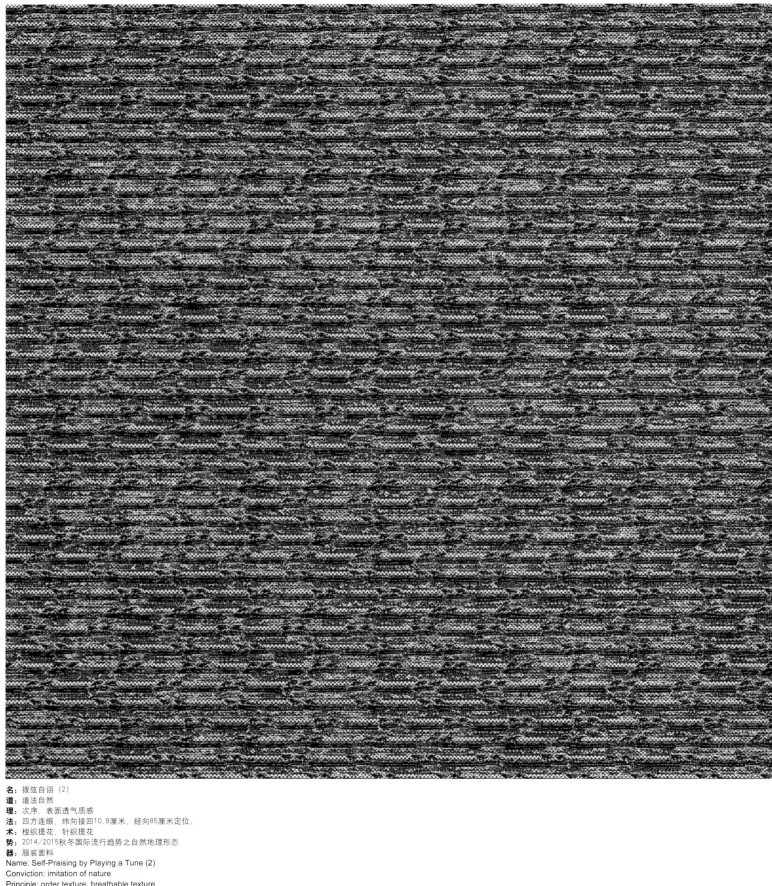

名：拨弦自诩（2）
道：道法自然
理：次序、表面透气质感
法：四方连缀、纬向接回10.8厘米，经向85厘米定位；
术：梭织提花，针织提花
势：2014/2015秋冬国际流行趋势之自然地理形态
器：服装面料

Name: Self-Praising by Playing a Tune (2)
Conviction: imitation of nature
Principle: order texture, breathable texture
Method: 4 sides in continuation, zonal tying-in 10.8 cm, longitudinal positioning 85 cm
Operation: jacquard weaving, jacquard knitting
Trend: 2014/2015 international trend of geo feature
Application: clothing fabrics

时间：2011年8月1日
地点：新疆塔什库尔干慕士塔格峰山麓
对象：积雪
灵感：高原上的积雪与植物
Time: August 1, 2011
Location: Foothills of Muztag Ata, Taxkorgan, Xinjiang
Object: the snow
Inspiration: snow and plant on the plateau

Creative Fabrics Design 1

名：拨弦自诩（7）、（9）
道：天工开物
理：次序、表面吸潮质感
法：四方连缀、纬向接回10.8厘米，经向85厘米定位
术：梭织提花、针织提花
势：2014/2015秋冬国际流行趋势之民族文化形态
器：服装面料

Name: Self-Praising by Playing a Tune (7), (9)
Conviction: heavenly creations
Principle: order texture, moisture absorption surface
Method: 4 sides in continuation, zonal tying-in 10.8 cm, longitudinal positioning 85 cm
Operation: jacquard weaving, jacquard knitting
Trend: 2014/2015 international trend of national culture
Application: clothing fabrics

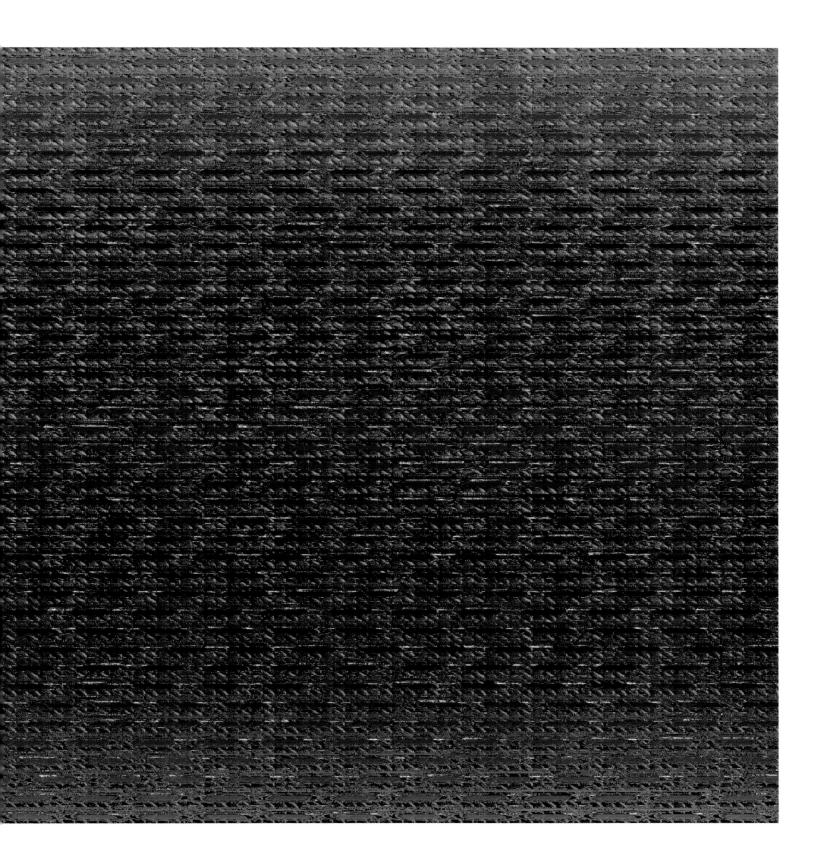

Creative Fabrics Design 1

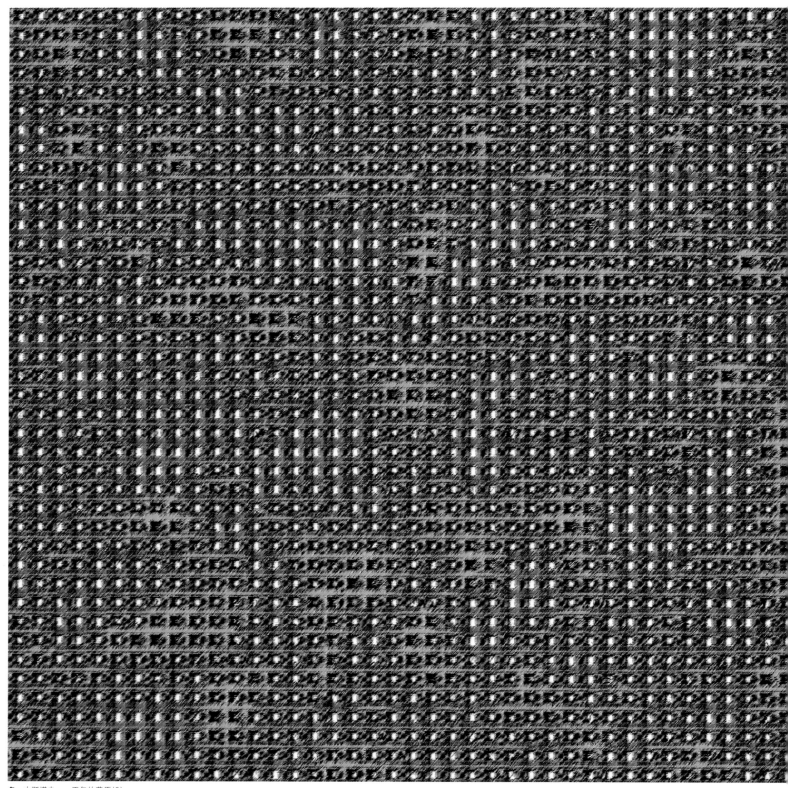

名：古斯塔夫——正午的草原(3)
道：道法自然、天工开物
理：次序、表面透气质感
法：四方连缀、纬向接回10.8厘米，经向85厘米定位
术：梭织提花，针织提花
势：2014/2015 秋冬国际流行趋势自然地理和民族文化形态
器：服装面料

Name: Gustaf — Midday Grassland (3)
Conviction: imitation of nature, heavenly creations
Principle: order texture, breathable texture
Method: 4 sides in continuation, zonal tying-in 10.8 cm, longitudinal positioning positioning 85 cm
Operation: jacquard weaving, jacquard knitting
Trend: 2014/2015 international trend of geo feature and national culture
Application: clothing fabrics

创意面料设计 第一卷

时间：2009年1月27日
地点：上海浦东金茂大厦
对象：玻璃幕墙
灵感：次序、光感
Time: January 27, 2009
Location: Jinmao Tower, Pudong, Shanghai
Object: glass curtain wall
Inspiration: order and light

Creative Fabrics Design 1

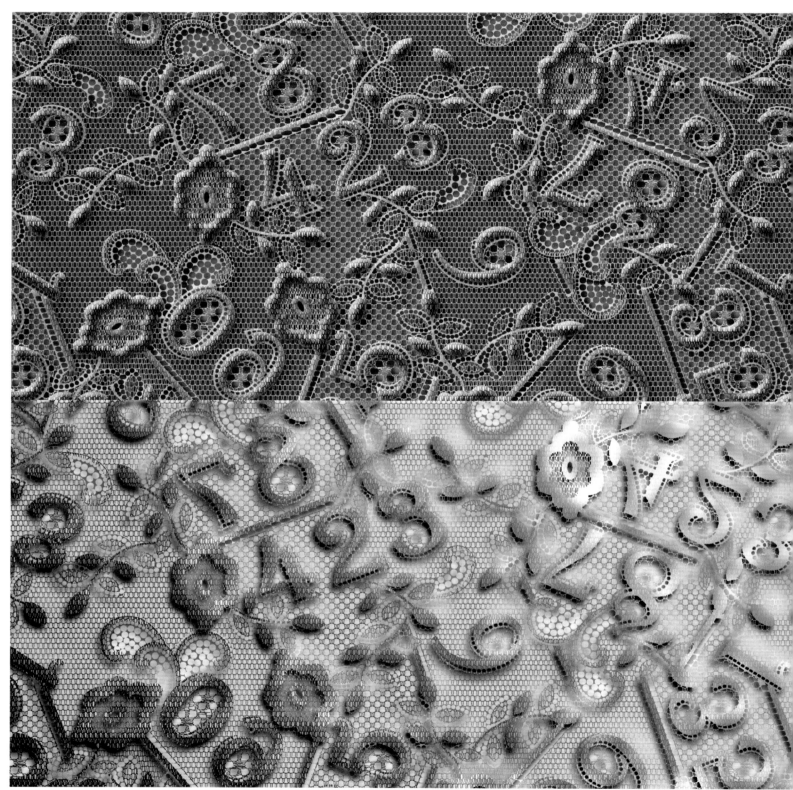

名：蕾丝——老数字（2）、（3）
道：天工开物
理：次序，表面透气质感
法：四方连缀
术：梭织提花，针织蕾丝
势：2014/2015秋冬国际流行趋势之造型形态
器：服装面料

Name: Lace — Old Figures (2), (3)
Conviction: heavenly creations
Principle: order texture, breathable texture
Method: 4 sides in continuation
Operation: jacquard weaving, lace knitting
Trend: 2014/2015 autumn and winter international fashion trend of styling
Application: clothing fabrics

时间：2009年12月5日
地点：浙江溪口
对象：阳光下的树影
灵感：层次、光感
Time: December 5, 2009
Location: Xikou, Zhejiang
Object: shadows of trees under the sun
Inspiration: shade and light

Creative Fabrics Design 1

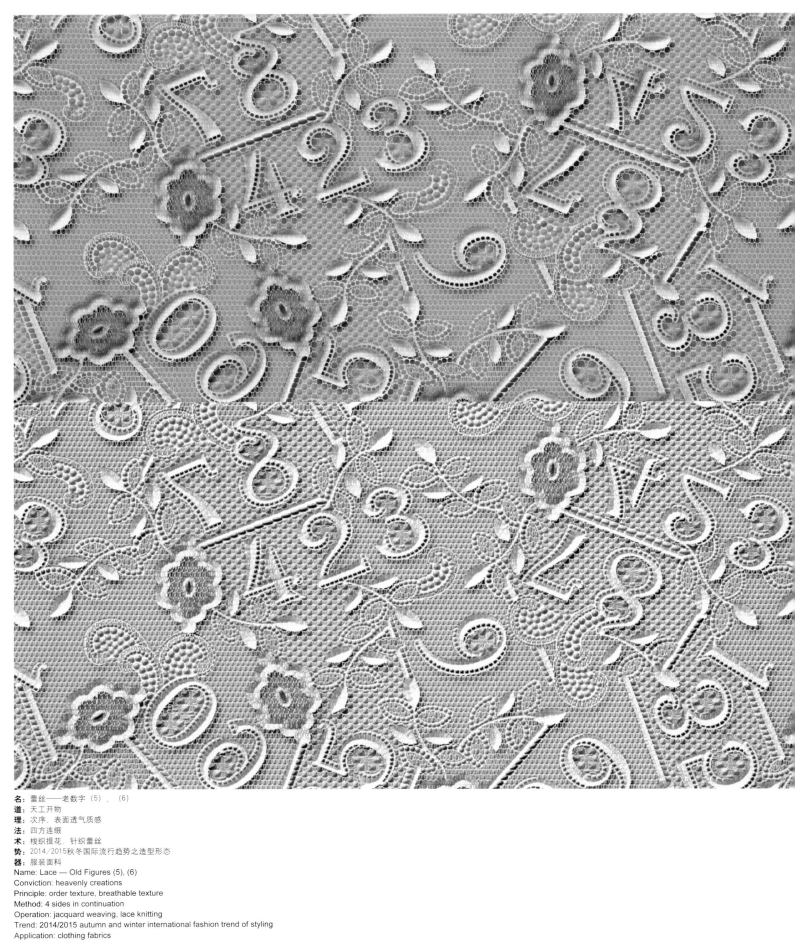

名：蕾丝——老数字（5）、(6)
道：天工开物
理：次序、表面透气质感
法：四方连缀
术：梭织提花、针织蕾丝
势：2014/2015秋冬国际流行趋势之造型形态
器：服装面料

Name: Lace — Old Figures (5), (6)
Conviction: heavenly creations
Principle: order texture, breathable texture
Method: 4 sides in continuation
Operation: jacquard weaving, lace knitting
Trend: 2014/2015 autumn and winter international fashion trend of styling
Application: clothing fabrics

时间：2009年12月4日
地点：浙江溪口蒋氏故居
对象：花格玻璃窗
灵感：刻花、光感
Time: December 4, 2009
Location: Jiang's Former Residence, Xikou, Zhejiang
Object: grillwork glass window
Inspiration: carving and glossing

Creative Fabrics Design 1

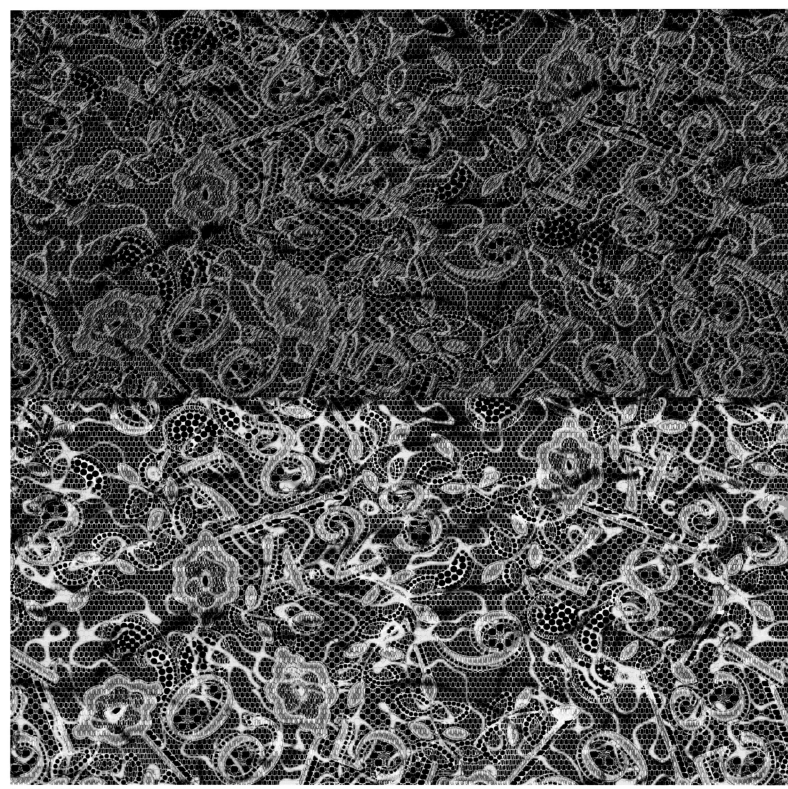

名：蕾丝——老数字（8）、（9）
道：天工开物
理：次序，表面透气质感
法：四方连缀
术：梭织提花，针织蕾丝
势：2014/2015秋冬国际流行趋势之造型形态
器：服装面料

Name: Lace — Old Figures (8), (9)
Conviction: heavenly creations
Principle: order texture, breathable texture
Method: 4 sides in continuation
Operation: jacquard weaving, lace knitting
Trend: 2014/2015 autumn and winter international fashion trend of styling
Application: clothing fabrics

时间：2012年7月30日
地点：四川成都锦里
对象：月下树影
灵感：剪影、摇曳生姿
Time: July 30, 2012
Location: Jinli Street, Chengdu, Sichuan
Object: shadow of trees under the moon
Inspiration: silhouette, swaying

Creative Fabrics Design 1

名：人迹罕至——冰川（1）
道：道法自然
理：堆积、滋长、表面透气质感
法：四方连缀
术：梭织提花，针织提花
势：2014/2015秋冬国际流行趋势之造型形态
器：服装面料，装饰布

Name: Virgin Land — the Glacier (1)
Conviction: imitation of nature
Principle: accumulation, growth, ventilation texture
Method: 4 sides in continuation
Operation: jacquard weaving, jacquard knitting
Trend: 2014/2015 autumn and winter international fashion trend of styling
Application: clothing fabrics, decorative cloth

时间：2009年7月21日
地点：青海玉珠峰
对象：南坡冰川
灵感：固体波浪
Time: July 21, 2009
Location: Yuzhu Peak, Qinghai
Object: glaceier in the south slope
Inspiration: solid wave

Creative Fabrics Design 1

名：人迹罕至——向高迪致敬（1）（局部）
道：道法自然
理：植物生长纹理
法：四方连缀，纬向接回35厘米，经向定位85厘米
术：绉纱组织、腐蚀
势：2014/2015国际流行趋势之自然生物形态
器：服装面料、装饰布

Name: Virgin Land — A Tribute to Gaudi (1) (portion)
Conviction: imitation of nature
Principle: texture of plant growth
Method: 4 sides in continuation, zonal tying-in 35 cm, longitudinal posstioning 85 cm
Operation: crepe silk, corroding
Trend: 2014/2015 international fashion trend of bio feature
Application: clothing fabrics, decorative cloth

时间：2010年8月7日
地点：云南丽江束河
对象：树干上的附生植物
灵感：植物重复生长肌理对比
Time: August 7, 2010
Location: Shuhe, Lijiang, Yunnan
Object: epiphytes on the trunk
Inspiration: texture contrast of plant growth in repetition

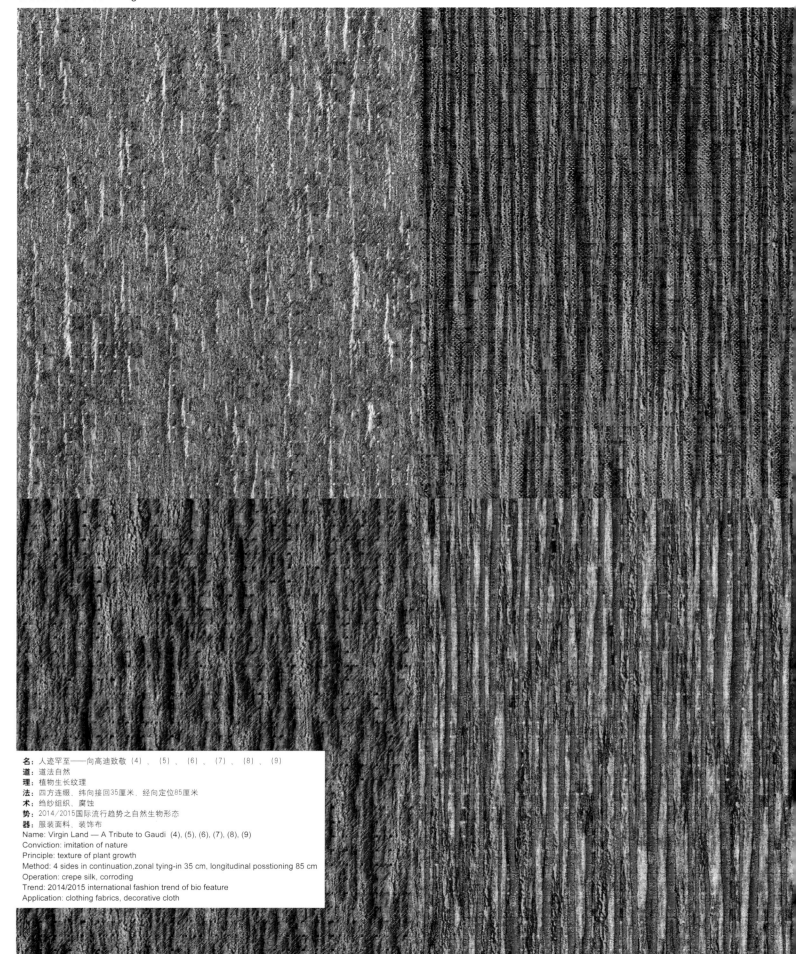

名：人迹罕至——向高迪致敬（4）、（5）、（6）、（7）、（8）、（9）
道：道法自然
理：植物生长纹理
法：四方连缀、纬向接回35厘米、经向定位85厘米
术：绉纱组织、腐蚀
势：2014/2015国际流行趋势之自然生物形态
器：服装面料、装饰布

Name: Virgin Land — A Tribute to Gaudi (4), (5), (6), (7), (8), (9)
Conviction: imitation of nature
Principle: texture of plant growth
Method: 4 sides in continuation, zonal tying-in 35 cm, longitudinal posstioning 85 cm
Operation: crepe silk, corroding
Trend: 2014/2015 international fashion trend of bio feature
Application: clothing fabrics, decorative cloth

Creative Fabrics Design 1

名：坐而论道——针织上的镶嵌心环纹（1）、（2）、（3）
道：民族文化
理：建筑物装饰镶嵌肌理
法：四方连缀、纬向接回35厘米、经向定位85厘米
术：绉纱组织、腐蚀、发泡
势：2014/2015国际流行趋势之民族文化形态
器：服装面料、装饰布

Name: Seated to Discuss the Tao — Inlaid Heart-shaped Ring Pattern in Knitting (1), (2), (3)
Conviction: national culture
Principle: decorative inlaid texture of architecture
Method: 4 sides in continuation, zonal tying-in 35 cm, longitudinal positioning 85 cm
Operation: crepe texture, corroding, foaming
Trend: 2014/2015 international fashion trend of national culture
Application: clothing fabrics, decorative cloth

Creative Fabrics Design 1

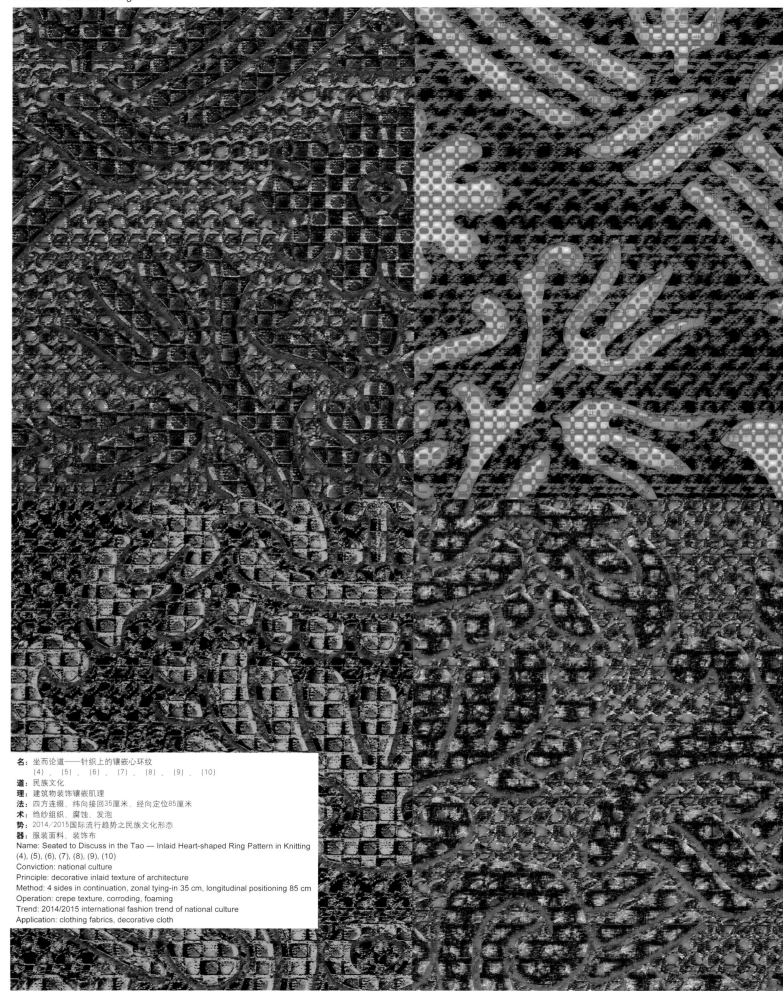

名：坐而论道——针织上的镶嵌心环纹
(4)、(5)、(6)、(7)、(8)、(9)、(10)
道：民族文化
理：建筑物装饰镶嵌肌理
法：四方连缀、纬向接回35厘米，经向定位85厘米
术：绉纱组织、腐蚀、发泡
势：2014/2015国际流行趋势之民族文化形态
器：服装面料、装饰布

Name: Seated to Discuss in the Tao — Inlaid Heart-shaped Ring Pattern in Knitting (4), (5), (6), (7), (8), (9), (10)
Conviction: national culture
Principle: decorative inlaid texture of architecture
Method: 4 sides in continuation, zonal tying-in 35 cm, longitudinal positioning 85 cm
Operation: crepe texture, corroding, foaming
Trend: 2014/2015 international fashion trend of national culture
Application: clothing fabrics, decorative cloth

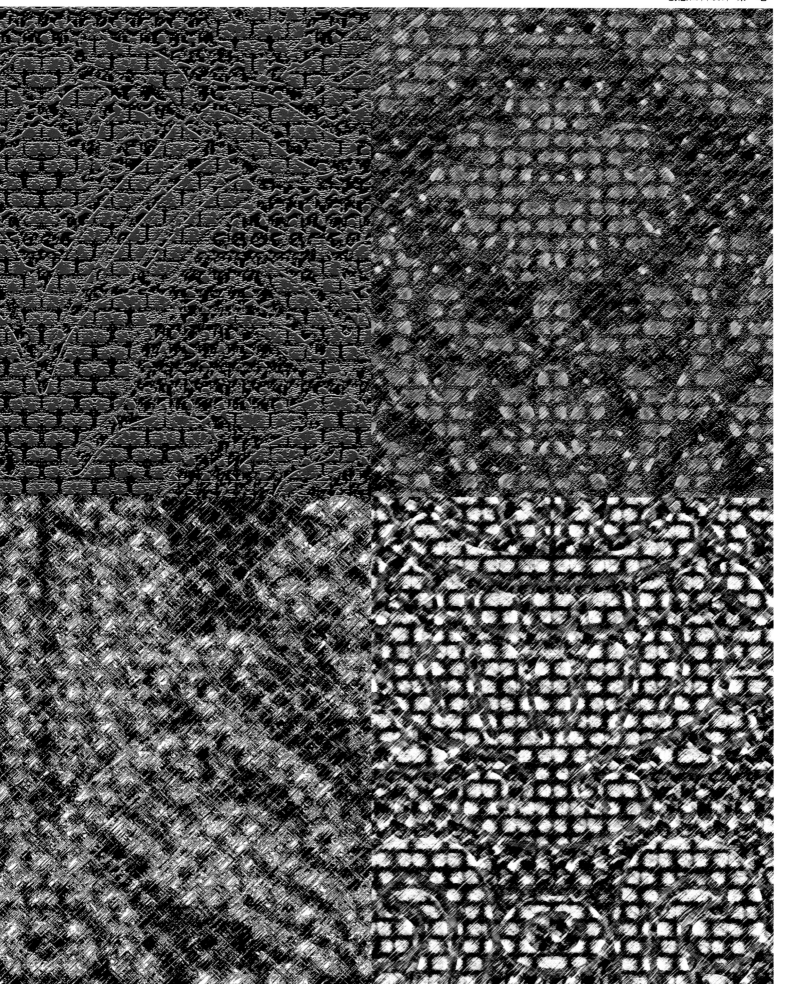

Creative Fabrics Design 1

名：坐而论道——针织上的镶嵌心环纹（11）（局部）
道：宗教艺术
理：釉色砖镶嵌
法：二方连缀，纬向接回35厘米，经向定位85厘米
术：针织组织，梭织绉纱组织，轧花、腐蚀
势：2014/2015国际流行趋势之民族文化形态
器：服装面料，装饰布
Name: Seated to Discuss the Tao — Inlaid Heart-shaped Ring Pattern in Knitting (11) (portion)
Conviction: religious art
Principle: glaze mosaic brick
Method: 2 sides in continuation, zonal tying-in 35 cm, longitudinal positioning 85 cm
Operation: knitting, weaving crepe texture, embossing corroding
Trend: 2014/2015 international fashion trend of national culture
Application: clothing fabrics, decorative cloth

时间：2011年7月30日
地点：新疆喀什香妃墓
对象：伊斯兰花式瓷砖贴面
灵感：纹样近似、釉色斑斓
Time: July 30, 2011
Location: Xiangfei Tomb, Xinjiang
Object: Islamic fancy tile
Inspiration: patterns, color gorgeous approximation

Creative Fabrics Design 1

名：坐而论道——针织上的镶嵌心环纹（5）
道：民族文化
理：堆积、对称
法：四方连缀
术：梭织提花，针织提花
势：2014/2015秋冬国际流行趋势之造型形态
器：服装面料、装饰布
Name: Seated to Discuss the Tao — Inlaid Heart-shaped Ring Pattern in Knitting (5)
Conviction: national culture
Principle: accumulation, symmetry
Method: 4 sides in continuation
Operation: woven jacquard, jacquard knitting
Trend: 2014/2015 autumn and winter international fashion trend of styling
Application: clothing fabrics, decorative cloth

时间：2012年8月7日
地点：云南中甸
对象：纺织品，建筑纹饰
灵感：五彩缤纷
Time: August 7, 2012
Location: Zhongdian, Yunnan
Object: textile, building decoration
Inspiration: a riot of colors

Creative Fabrics Design 1

名： 针织——奥菲利亚的晚唱 (1)、(2)、(3)、(4)、(5)、(6)、(7)、(8)
道： 道法自然
理： 排列、近似、滋长、织物肌理
法： 针织组织、梭织二方连缀、纬向接回21.6厘米、经向定位85厘米
术： 针织组织、梭织绉纱组织、轧花、腐蚀
势： 2014/2015国际流行趋势之民族文化形态
器： 服装面料、装饰布

Name: Knitting — Ophelia's Nocturne (1), (2), (3), (4), (5), (6), (7), (8)
Conviction: imitation of nature
Principle: sequence, approximation, growing, fabric texture
Method: knitting, weaving 2 sides in continuation, zonal tying-in 21.6 cm, longitudinal positioning 85 cm
Operation: knitting, weaving crepe texture, embossing, corroding
Trend: 2014/2015 international fashion trend of national culture
Application: clothing fabrics, decorative cloth

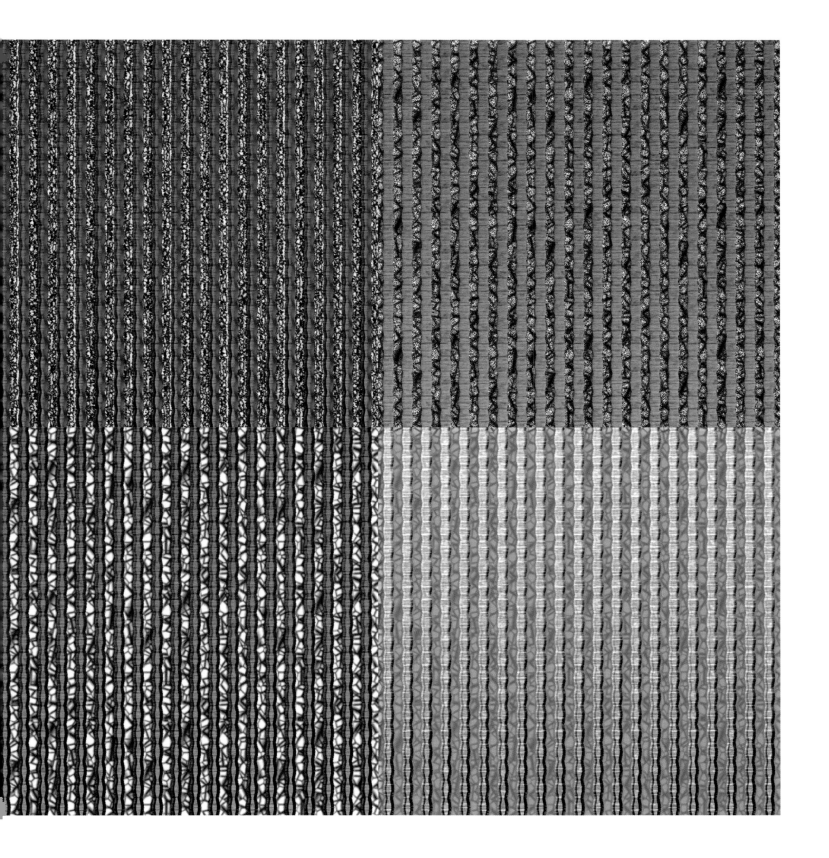

Creative Fabrics Design 1

名：针织——奥菲利亚的晚唱 (9)、(10)、(11)、(12)、(13)、(14)、(15)、(16)
道：道法自然
理：排列、近似、溢长、织物肌理
法：针织组织，梭织二方连缀，纬向接回21.6厘米，经向定位85厘米
术：针织组织，梭织绉纱组织、轧花、腐蚀
势：2014/2015国际流行趋势之民族文化形态
器：服装面料、装饰布

Name: Knitting — Ophelia's Nocturne (9), (10), (11), (12), (13), (14), (15), (16)
Conviction: imitation of nature
Principle: sequence, approximation, growing, fabric texture
Method: knitting, weaving 2 sides in continuation, zonal tying-in 21.6 cm, longitudinal positioning 85 cm
Operation: knitting, weaving crepe texture, embossing, corroding
Trend: 2014/2015 international fashion trend of national culture
Application: clothing fabrics, decorative cloth

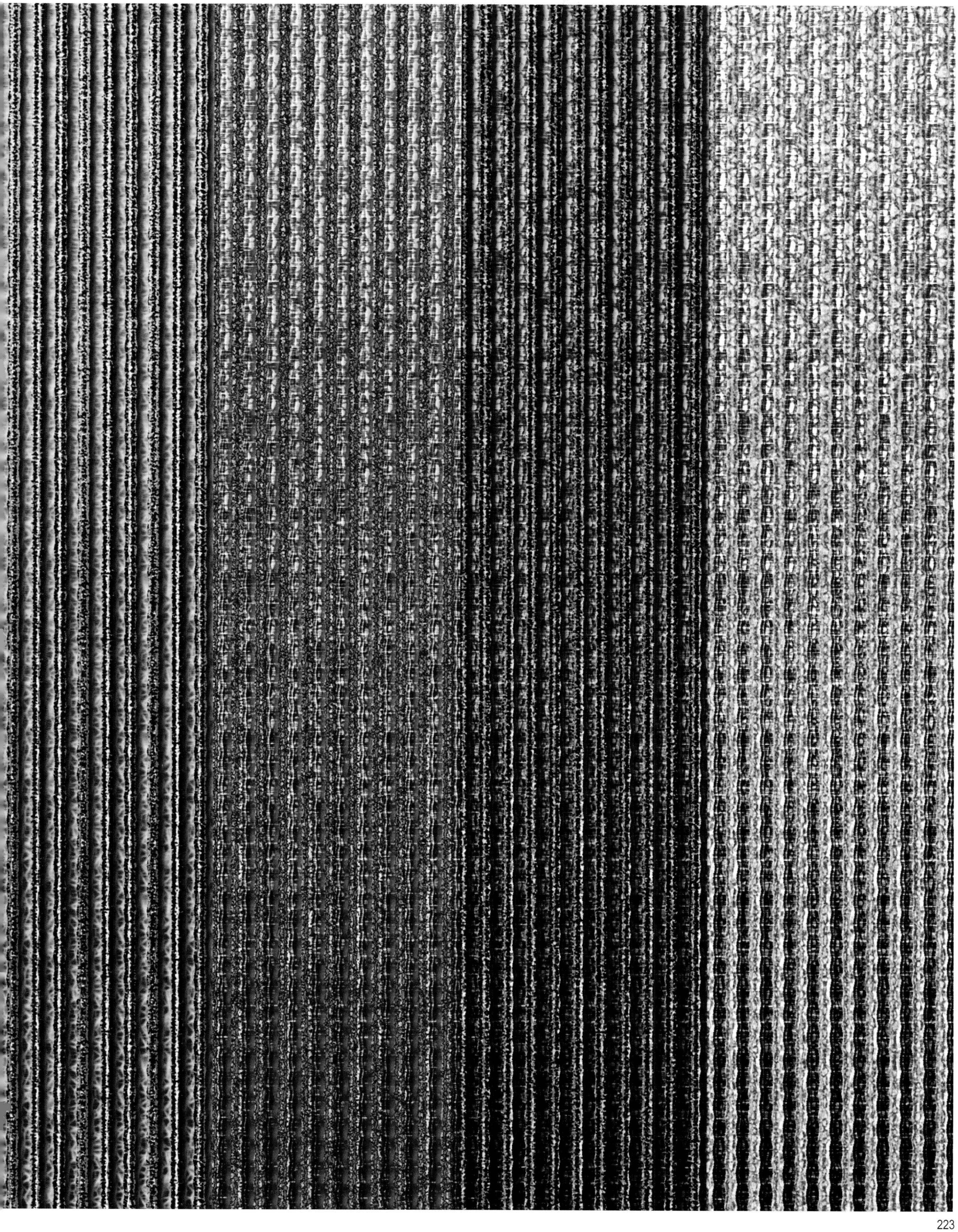

Creative Fabrics Design 1

名：针织——奥菲利亚的晚唱 (17)、(18)、(19)、(20)、(21)、(22)、(23)、(24)
道：道法自然
理：排列、近似、滋长、织物肌理
法：针织组织、梭织二方连缀、纬向接回21.6厘米、经向定位85厘米
术：针织组织、梭织绉纱组织、轧花、腐蚀
势：2014/2015国际流行趋势之民族文化形态
器：服装面料、装饰布
Name: Knitting — Ophelia's Nocturne (17), (18), (19), (20), (21), (22), (23), (24)
Conviction: imitation of nature
Principle: sequence, approximation, growing, fabric texture
Method: knitting, weaving 2 sides in continuation, zonal tying-in 21.6 cm, longitudinal positioning 85 cm
Operation: knitting, weaving crepe texture, embossing, corroding
Trend: 2014/2015 international fashion trend of national culture
Application: clothing fabrics, decorative cloth

Creative Fabrics Design 1

名：针织——奥菲利亚的晚唱（25）、（26）、（27）、（28）、（29）、（30）、（31）、（32）
道：道法自然
理：排列、近似、滋长、织物肌理
法：针织组织、梭织二方连缀、纬向接回21.6厘米、经向定位85厘米
术：针织组织、梭织绉纱组织、轧花、腐蚀
势：2014/2015国际流行趋势之民族文化形态
器：服装面料、装饰布

Name: Knitting — Ophelia's Nocturne (25), (26), (27), (28), (29), (30), (31), (32)
Conviction: imitation of nature
Principle: sequence, approximation, growing, fabric texture
Method: knitting, weaving 2 sides in continuation, zonal tying-in 21.6 cm, longitudinal positioning 85 cm
Operation: knitting, weaving crepe texture, embossing, corroding
Trend: 2014/2015 international fashion trend of national culture
Application: clothing fabrics, decorative cloth

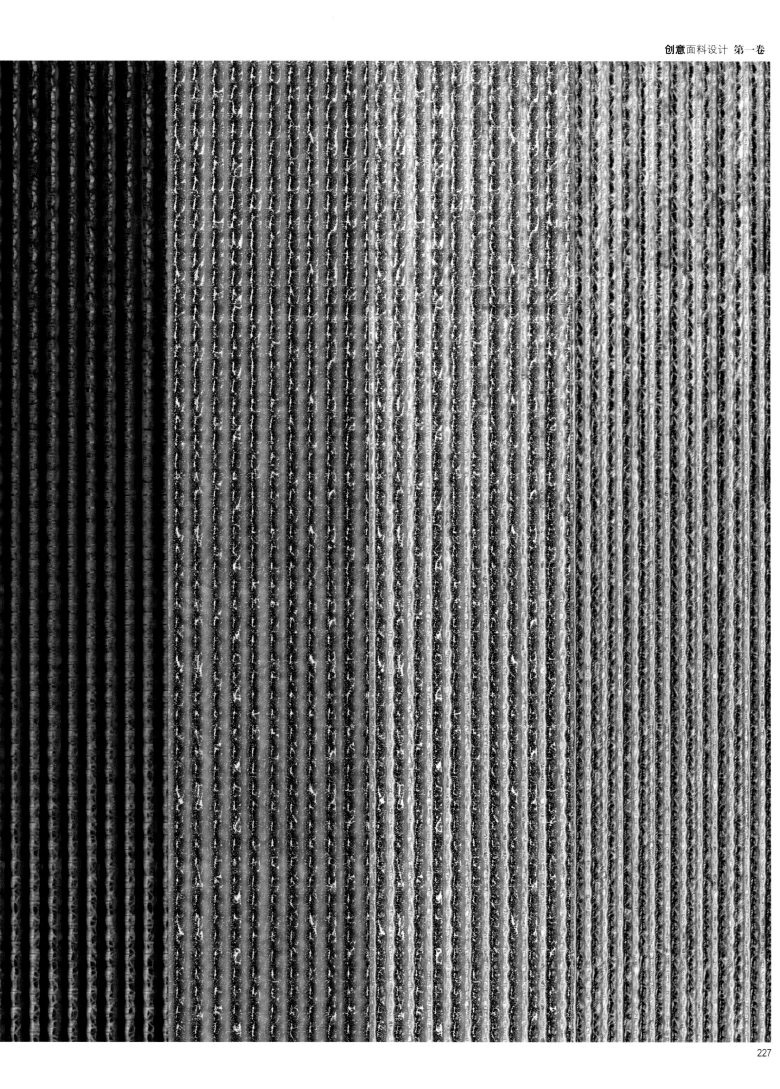

Creative Fabrics Design 1

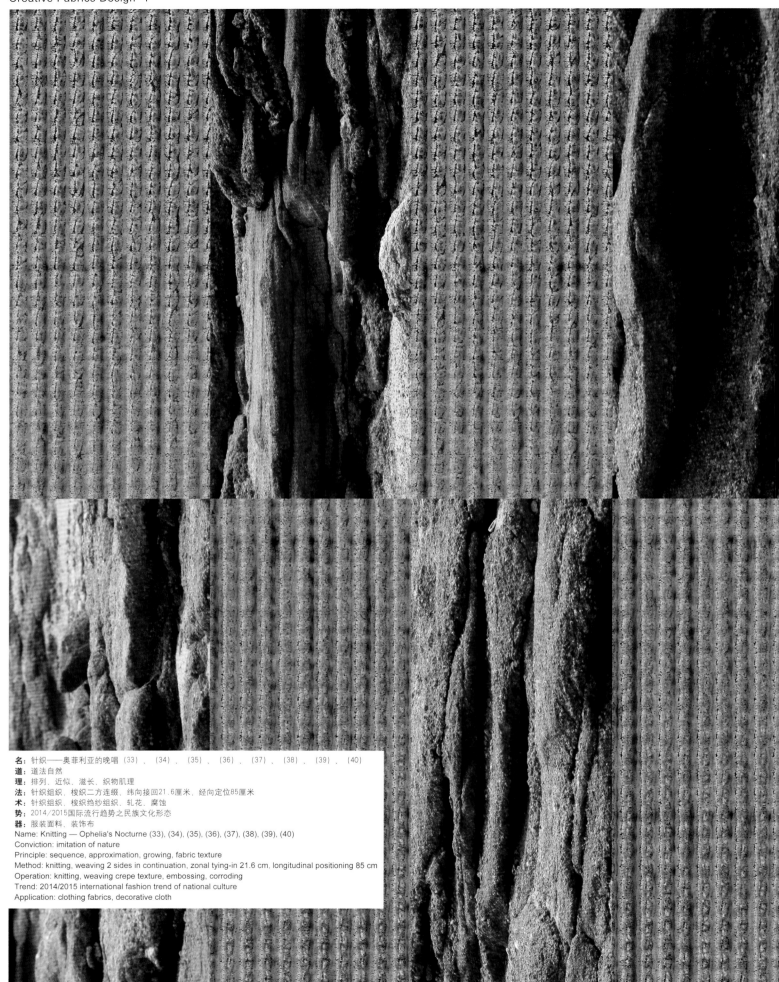

名：针织——奥菲利亚的晚唱 (33)、(34)、(35)、(36)、(37)、(38)、(39)、(40)
道：道法自然
理：排列、近似、滋长、织物肌理
法：针织组织，梭织二方连缀，纬向接回21.6厘米，经向定位85厘米
术：针织组织、梭织绉纱组织、轧花、腐蚀
势：2014/2015国际流行趋势之民族文化形态
器：服装面料、装饰布
Name: Knitting — Ophelia's Nocturne (33), (34), (35), (36), (37), (38), (39), (40)
Conviction: imitation of nature
Principle: sequence, approximation, growing, fabric texture
Method: knitting, weaving 2 sides in continuation, zonal tying-in 21.6 cm, longitudinal positioning 85 cm
Operation: knitting, weaving crepe texture, embossing, corroding
Trend: 2014/2015 international fashion trend of national culture
Application: clothing fabrics, decorative cloth

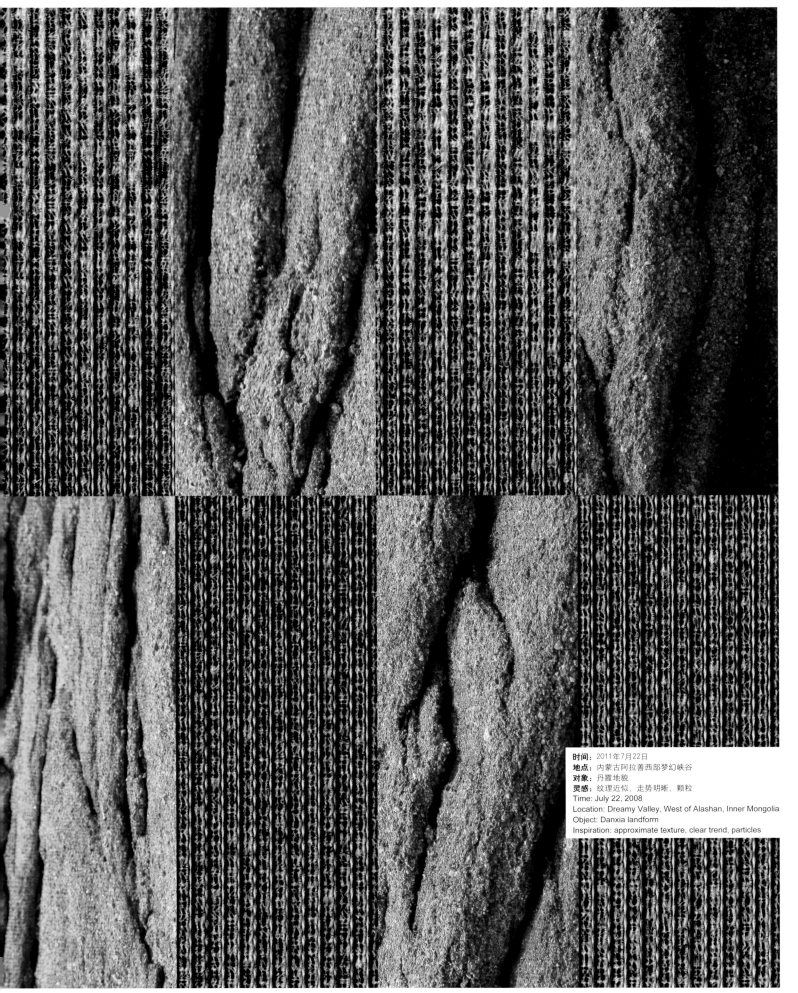

时间：2011年7月22日
地点：内蒙古阿拉善西部梦幻峡谷
对象：丹霞地貌
灵感：纹理近似、走势明晰、颗粒
Time: July 22, 2008
Location: Dreamy Valley, West of Alashan, Inner Mongolia
Object: Danxia landform
Inspiration: approximate texture, clear trend, particles

名： 针织——奥菲利亚的晚唱（41）、（42）、（43）
道： 道法自然
理： 排列、近似、滋长、织物肌理
法： 针织组织、梭织二方连缀、纬向接回21.6厘米，经向定位85厘米
术： 针织组织，梭织绉纱组织、轧花、腐蚀
势： 2014/2015国际流行趋势之民族文化形态
器： 服装面料、装饰布

Name: Knitting — Ophelia's Nocturne (41), (42), (43)
Conviction: imitation of nature
Principle: sequence, approximation, growing, fabric texture
Method: knitting, weaving 2 sides in continuation, zonal tying-in 21.6 cm, longitudinal positioning 85 cm
Operation: knitting, weaving crepe texture, embossing, corroding
Trend: 2014/2015 international fashion trend of national culture
Application: clothing fabrics, decorative cloth

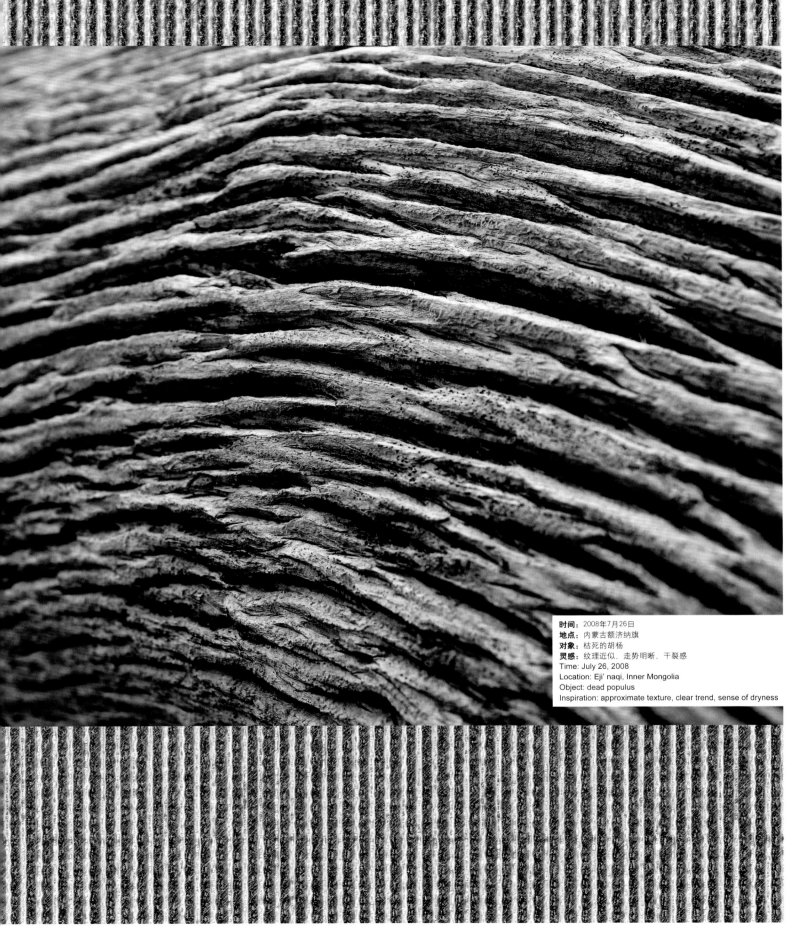

时间：2008年7月26日
地点：内蒙古额济纳旗
对象：枯死的胡杨
灵感：纹理近似、走势明晰、干裂感
Time: July 26, 2008
Location: Eji' naqi, Inner Mongolia
Object: dead populus
Inspiration: approximate texture, clear trend, sense of dryness

Creative Fabrics Design 1

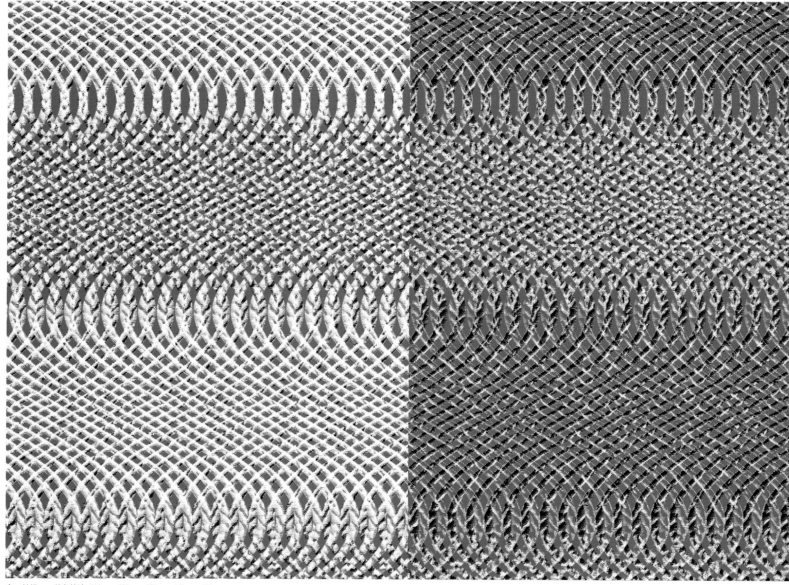

名：针织——慧泉悖论（1）、（2）、（3）、（4）、（5）
道：数字逻辑
理：排列、近似、滋长、织物肌理
法：针织组织、梭织二方连缀、纬向接回21.6厘米、经向定位85厘米
术：针织组织、梭织绉纱组织、轧花、腐蚀
势：2014/2015国际流行趋势之民族文化形态
器：服装面料、装饰布

Name: Knitting — Paradox of the Wisdom Spring (1), (2), (3), (4), (5)
Conviction: digital logic
Principle: sequence, approximation, growing, fabric texture
Method: knitting, weaving 2 sides in continuation, zonal tying-in 21.6 cm, longitudinal positioning 85 cm
Operation: knitting, weaving crepe texture, embossing, corroding
Trend: 2014/2015 international fashion trend of national culture
Application: clothing fabrics, decorative cloth

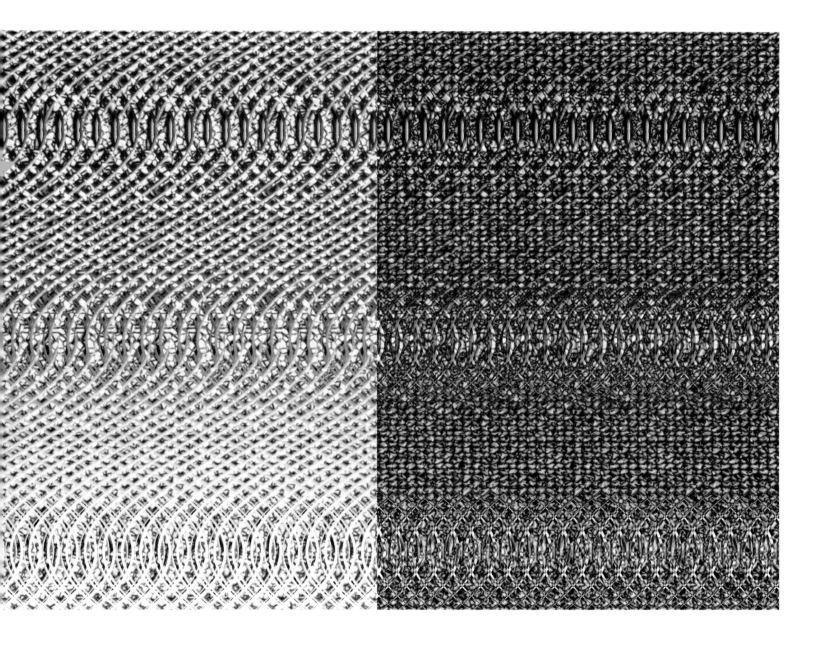

Creative Fabrics Design 1

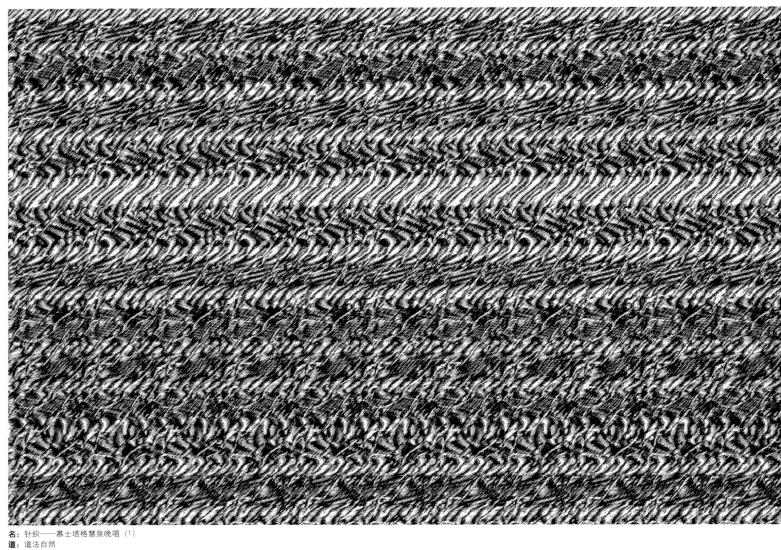

名：针织——慕士塔格慧泉晚唱（1）
道：道法自然
理：堆积、滋长、风砺、动静相宜
法：针织组织、梭织二方连缀、纬向接回21.6厘米、经向定位85厘米
术：梭织提花、针织提花
势：2014/2015秋冬国际流行趋势之造型形态
器：服装面料、装饰布
Name: Knitting — Singing of the Wisdom Spring at Muztag Ata (1)
Conviction: heavenly creations
Principle: accumulation, growing wind effect, balance between action and silence
Method: weaving 2 sides in continuation, zonal tying-in 21.6 cm, longitudinal positioning 85 cm
Operation: jacquard woven, jacquard knitting
Trend: 2014/2015 international fashion trend of styling
Application: clothing fabrics, decorative cloth

时间：2011年8月1日
地点：新疆塔什库尔干慕士塔格雪山
对象：冰川雪水
灵感：冰山上的来客
Time: August 1, 2011
Location: Muztag Ata, Taxkorgan, Xinjiang
Object: snow water from glacier
Inspiration: visitors from the ice-capped mountain

Creative Fabrics Design 1

名：世博旋律唱晚（1）
道：天工开物
理：数字次序
法：针织组织、梭织二方连缀、纬向接回21.6厘米、经向定位85厘米
术：针织组织、梭织绉纱组织、轧花、腐蚀
势：2014/2015国际流行趋势之科技形态
器：服装面料、装饰布

Name: The Melody of World Expo (1)
Conviction: heavenly creations
Principle: numerical order
Method: knitting, weaving 2 sides in continuation, zonal tying-in 21.6 cm, longitudinal positioning 85 cm
Operation: knitting, weaving crepe texture, embossing, corroding
Trend: 2014/2015 international fashion trend of science and technology
Application: clothing fabrics, decorative cloth

名：针织——搏杀青砖（1）
道：天工开物
理：堆积聚众、近似次序、陈年风化
法：针织组织，梭织二方连缀、纬向接回21.6厘米，经向定位85厘米
术：梭织提花，针织提花
势：2014/2015秋冬国际流行趋势之造型形态
器：服装面料、装饰布

Name: Knitting — the Traditional Black Bricks (1)
Conviction: heavenly creations
Principle: accumulation, approximation order, weathering
Method: knitting, weaving 2 sides in continuation, zonal tying-in 21.6 cm, longitudinal positioning 85 cm
Operation: jacquard weaving, jacquard knitting
Trend: 2014/2015 autumn and winter international fashion trend of styling
Application: clothing fabrics, decorative cloth

时间：2010年7月31日
地点：四川成都宽窄巷子
对象：青砖墙
灵感：陈年故居
Time: July 31, 2010
Location: Broad and Narrow Alley, Chengdu, Sichuan
Object: the black brick wall
Inspiration: old residence

Creative Fabrics Design 1

名：针织——异域镜像（1）
道：天工开物
理：堆积聚众、近似次序、陈年风化
法：针织组织，梭织二方连缀、纬向接回21.6厘米，经向定位85厘米
术：梭织提花，针织提花
势：2014/2015秋冬国际流行趋势之造型形态
器：服装面料、装饰布

Name: Knitting — the Exotic Mirror Image
Conviction: heavenly creations
Principle: accumulation, approximation order, weathering
Method: knitting, weaving 2 sides in continuation, zonal tying-in 21.6 cm, longitudinal positioning 85 cm
Operation: jacquard weaving, jacquard knitting
Trend: 2014/2015 autumn and winter international fashion trend of styling
Application: clothing fabrics, decorative cloth

时间：2011年7月30日
地点：新疆喀什
对象：建筑玻璃立面
灵感：镜像
Time: July 30, 2011
Location: Kashi, Xinjiang
Object: architectural glass facade
Inspiration: mirror image

Creative Fabrics Design 1

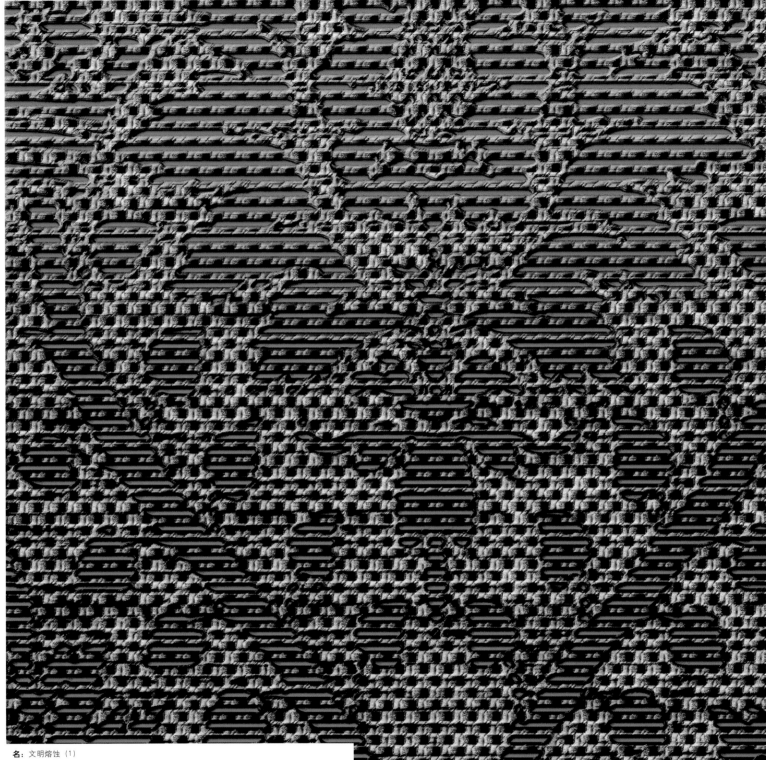

名：文明熔蚀（1）
道：天工开物
理：针织物数字次序
法：梭织二方连缀、纬向接回21.6厘米，经向定位85厘米
术：针织组织，梭织绉纱组织、轧花、腐蚀
势：2014/2015国际流行趋势之民族文化形态
器：服装面料、装饰布

Name: Civilization in Erosion (1)
Conviction: heavenly creations
Principle: knitted numerical order
Method: weaving 2 sides in continuation, zonal tying-in 21.6 cm, longitudinal positioning 85 cm
Operation: knitting, weaving crepe texture, embossing, corroding
Trend: 2014/2015 international fashion trend of national culture
Application: clothing fabrics, decorative cloth

时间：2011年7月30日
地点：新疆喀什香妃墓
对象：伊斯兰花式土砖
灵感：搭积木，次序
Time: July 30, 2011
Location: Xiangfei Tomb, Kashi, Xinjiang
Object: Islamic fancy brick
Inspiration: building blocks, order

Creative Fabrics Design 1

名：文明熔蚀（2）、（3）、（4）、（5）
道：天工开物
理：针织物数字次序
法：梭织二方连缀，纬向接回21.6厘米，经向定位85厘米
术：针织组织，梭织绉纱组织，轧花，腐蚀
势：2014/2015国际流行趋势之民族文化形态
器：服装面料，装饰布

Name: Civilization in Erosio (2), (3), (4), (5)
Conviction: heavenly creations
Principle: knitted numerical order
Method: weaving 2 sides in continuation, zonal tying-in 21.6 cm, longitudinal positioning 85 cm
Operation: knitting, weaving crepe texture, embossing, corroding
Trend: 2014/2015 international fashion trend of national culture
Application: clothing fabrics, decorative cloth

时间：2011年7z月30日
地点：新疆喀什香妃墓
对象：伊斯兰花式土砖
灵感：搭积木，次序
Time: July 30, 2011
Location: Xiangfei Tomb, Kashi, Xinjiang
Object: Islamic fancy brick
Inspiration: building blocks, order

Creative Fabrics Design 1

名：文明熔蚀 (6)、(7)、(10)、(11)
道：天工开物
理：针织物数字次序
法：梭织二方连缀，纬向接回21.6厘米，经向定位85厘米
术：针织组织，梭织绉纱组织、轧花、腐蚀
势：2014/2015国际流行趋势之民族文化形态
器：服装面料、装饰布

Name: Civilization in Erosion (6), (7), (10), (11)
Conviction: heavenly creations
Principle: knitted numerical order
Method: weaving 2 sides in continuation, zonal tying-in 21.6 cm, longitudinal positioning 85 cm
Operation: knitting, weaving crepe texture, embossing, corroding
Trend: 2014/2015 international fashion trend of national culture
Application: clothing fabrics, decorative cloth

名：文明熔蚀（8）、（9）
道：天工开物
理：针织物数字次序
法：梭织二方连缀，纬向接回21.6厘米，经向定位85厘米
术：针织组织，梭织绉纱组织、轧花、腐蚀
势：2014/2015国际流行趋势之民族文化形态
器：服装面料、装饰布

Name: Civilization in Erosion (8), (9)
Conviction: heavenly creations
Principle: knitted fabric numerical order
Method: weaving 2 sides in continuation, zonal tying-in 21.6 cm, longitudinal positioning 85 cm
Operation: knitting, weaving crepe texture, embossing, corroding
Trend: 2014/2015 international fashion trend of national culture
Application: clothing fabrics, decorative cloth

Creative Fabrics Design 1

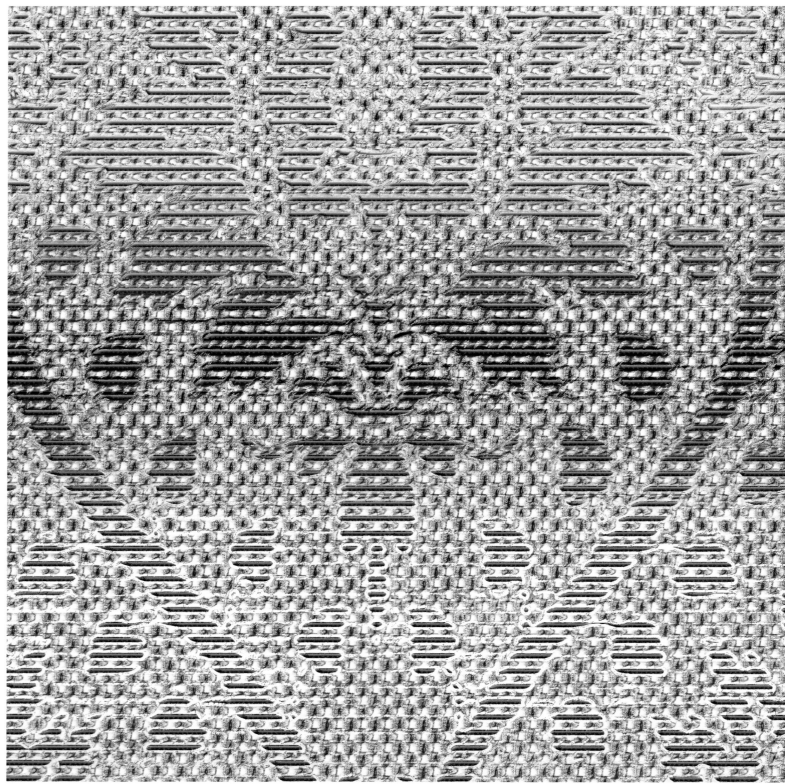

名：针织——文明熔蚀（12）
道：天工开物
理：针织物数字次序
法：梭织二方连缀、纬向接回21.6厘米，经向定位85厘米
术：针织组织，梭织绉纱组织、轧花、腐蚀
势：2014/2015国际流行趋势之民族文化形态
器：服装面料、装饰布；

Name: Knitting — Civilization in Erosion (12)
Conviction: heavenly creations
Principle: knitted fabric numerical order
Method: weaving 2 sides in continuation, zonal tying-in 21.6 cm, longitudinal positioning 85 cm
Operation: knitting, weaving crepe texture, embossing, corroding
Trend: 2014/2015 international fashion trend of national culture
Application: clothing fabrics, decorative cloth

时间：2009年8月4日
地点：四川甘孜
对象：风马旗
灵感：阵列

Time: August 4, 2009
Location: Ganzi, Sichuan
Object: the pray flags
Inspiration: array

Creative Fabrics Design 1

名：针织——文明熔蚀（13）、（16）
道：天工开物
理：针织物数字次序
法：梭织二方连缀，纬向接回21.6厘米，经向定位85厘米
术：针织组织，梭织绉纱组织，轧花，腐蚀
势：2014/2015国际流行趋势之民族文化形态
器：服装面料、装饰布

Name: Knitting — Civilization in Erosion (13), (16)
Conviction: heavenly creations
Principle: knitted numerical order
Method: weaving 2 sides in continuation, zonal tying-in 21.6 cm, longitudinal positioning 85 cm
Operation: knitting, weaving crepe texture, embossing, corroding
Trend: 2014/2015 international fashion trend of national culture
Application: clothing fabrics, decorative cloth

Creative Fabrics Design 1

名：菱形金枪头（1）、（2）
道：天工开物
理：几何次序、数字阵列
法：梭织二方连缀、纬向接回5厘米
术：针织组织、梭织绉纱组织、轧花、切割
势：2014/2015国际流行趋势之民族文化形态
器：服装面料、装饰布

Name: Golden Diamond-shape Spearhead (1), (2)
Conviction: heavenly creations
Principle: geometric sequence, digital array
Method: weaving 2 sides in continuation, zonal tying-in 5 cm
Operation: knitting, weaving crepe texture, embossing, cutting
Trend: 2014/2015 international fashion trend of national culture
Application: clothing fabrics, decorative cloth

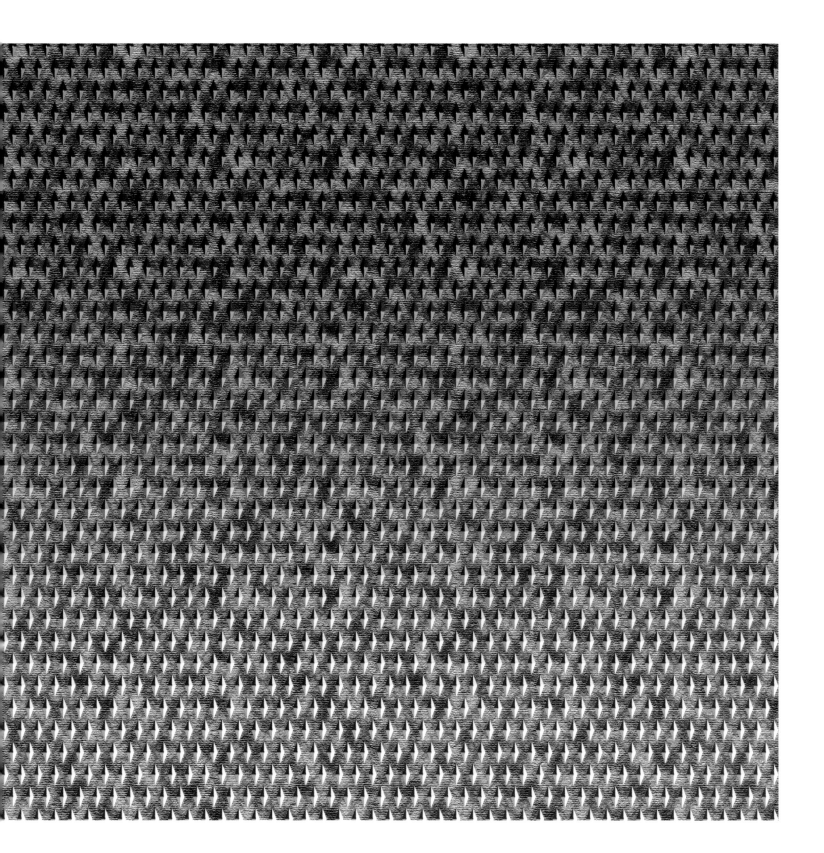

Creative Fabrics Design 1

名：菱形金枪头（3）、（4）、（5）、（6）、（7）、（8）、（9）
道：天工开物
理：几何次序、数字阵列
法：梭织二方连缀、纬向接回5厘米
术：针织组织、梭织绉纱组织、轧花、切割
势：2014/2015国际流行趋势之民族文化形态
器：服装面料、装饰布

Name: Golden Diamond-shape Spearhead
(3), (4), (5), (6), (7), (8), (9)
Conviction: heavenly creations
Principle: geometric sequence, digital array
Method: weaving 2 sides in continuation, zonal tying-in 5 cm
Operation: knitting, weaving crepe texture, embossing, cutting
Trend: 2014/2015 international fashion trend of national culture
Application: clothing fabrics, decorative cloth

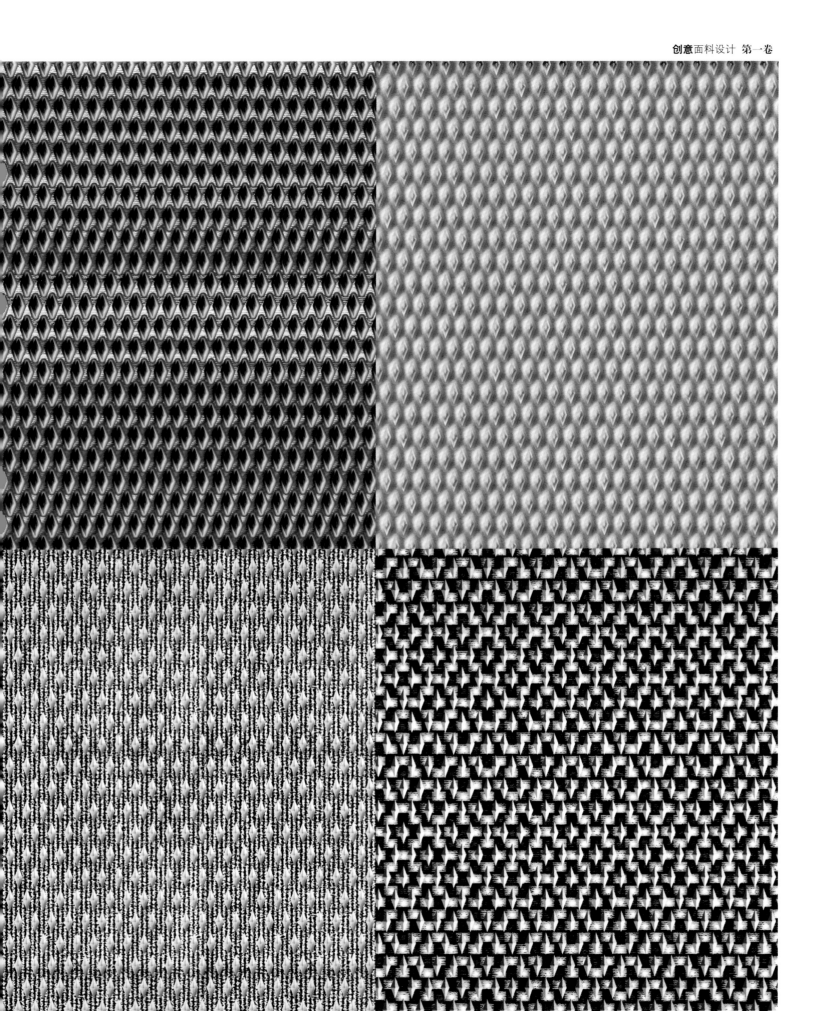

Creative Fabrics Design 1

名：菱形金枪头（10）、（11）、（12）、（13）
道：天工开物
理：几何次序、数字阵列
法：梭织二方连缀、纬向接回5厘米
术：针织组织、梭织绉纱组织、轧花、切割
势：2014/2015国际流行趋势之民族文化形态
器：服装面料、装饰布；

Name: Golden Diamond-shape Spearhead (10), (11), (12), (13)
Conviction: heavenly creations
Principle: geometric sequence, digital array
Method: weaving 2 sides in continuation, zonal tying-in 5 cm
Operation: knitting, weaving crepe texture, embossing, cutting
Trend: 2014/2015 international fashion trend of national culture
Application: clothing fabrics, decorative cloth

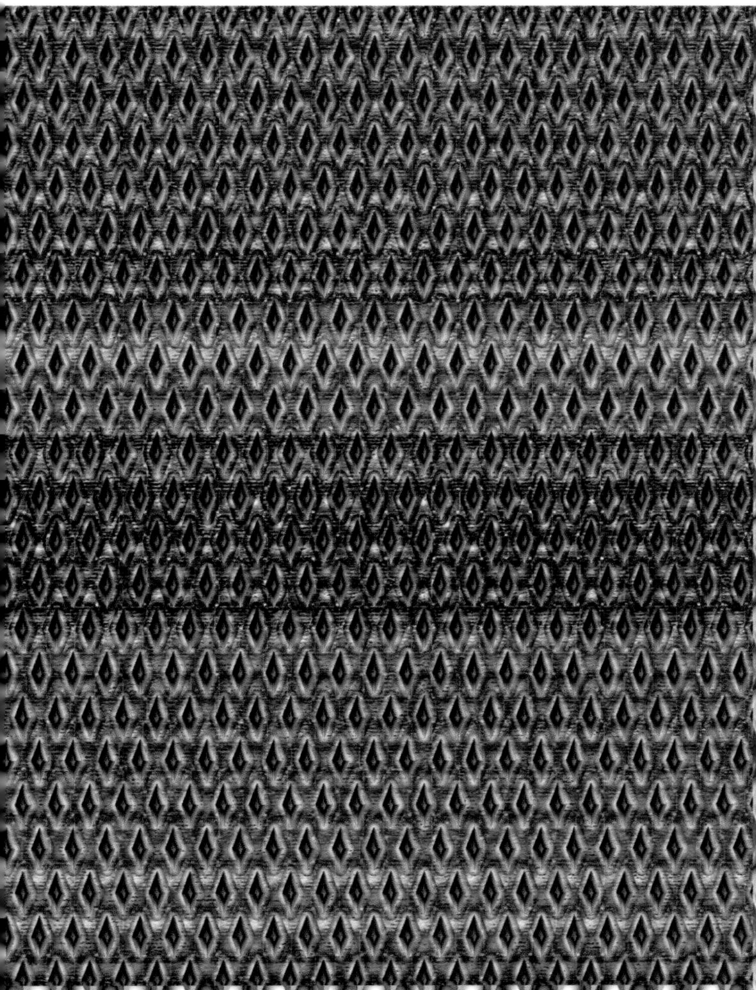

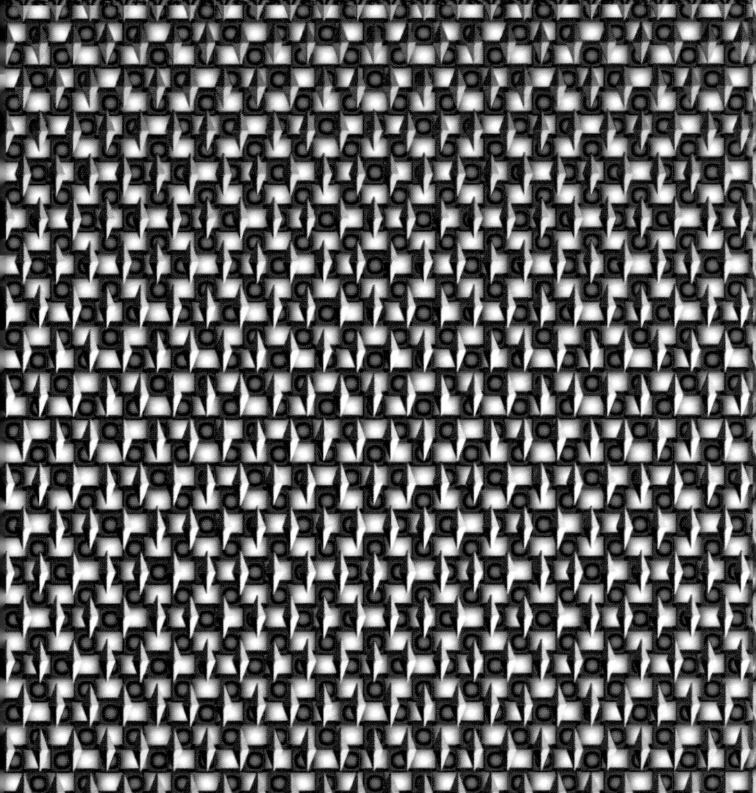

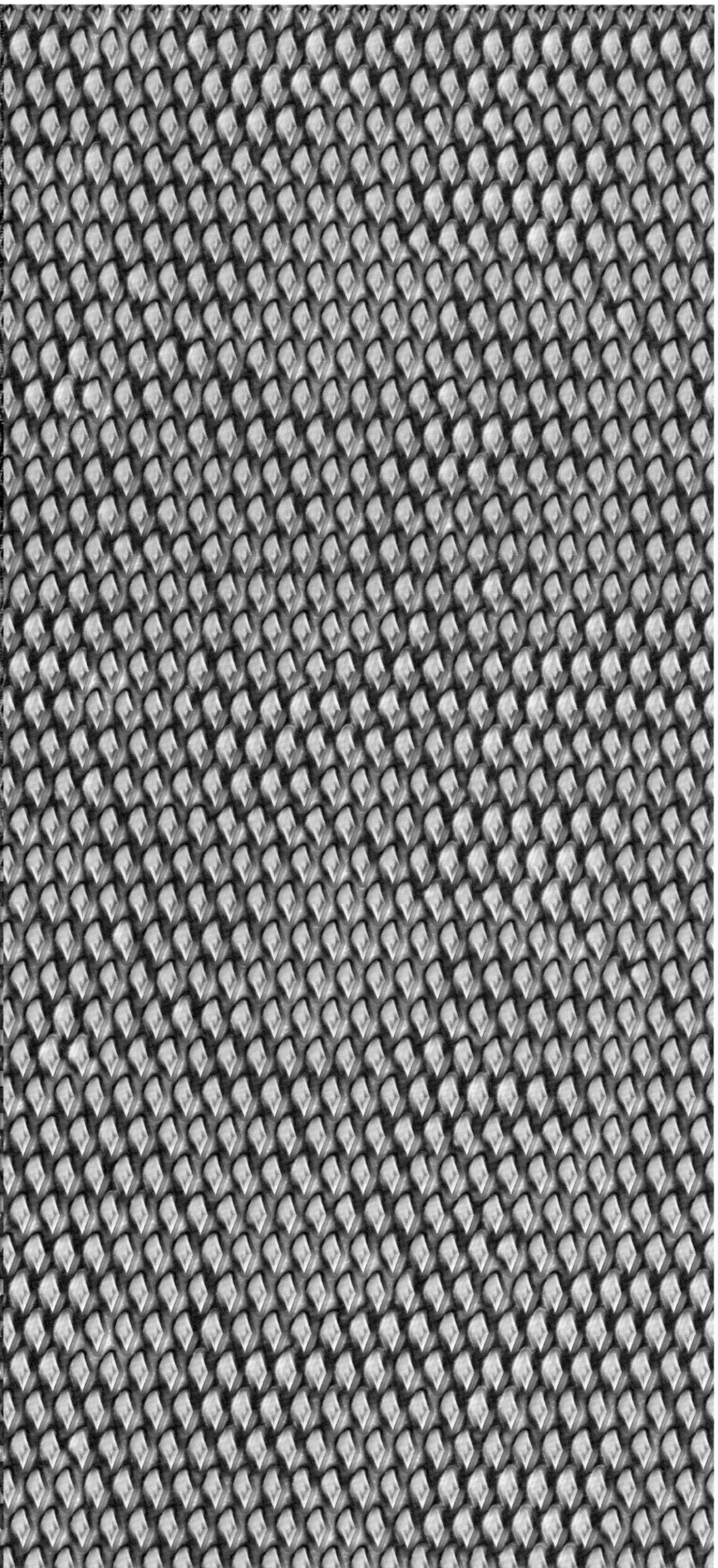

Creative Fabrics Design 1

名：菱形金枪头（16）
道：天工开物
理：几何次序、数字阵列
法：梭织二方连缀、纬向接回5厘米
术：针织组织、梭织绉纱组织、轧花、切割
势：2014/2015国际流行趋势之民族文化形态
器：服装面料、装饰布

Name: Golden Diamond-shape Spearhead (16)
Conviction: heavenly creations
Principle: geometric sequence, digital array
Method: weaving 2 sides in continuation, zonal tying-in 5 cm
Operation: knitting, weaving crepe texture, embossing, cutting
Trend: 2014/2015 international fashion trend of national culture
Application: clothing fabrics, decorative cloth

时间：2010年11月29日
地点：上海淮海东路卡地亚专卖店
对象：幕墙
灵感：阵列、光感、色感
Time: November 29, 2010
Location: Cartire stores of East Huaihai Road, Shanghai
Object: curtain wall
Inspiration: array, light, color

Creative Fabrics Design 1

名：菱形金枪头（17）、（18）、（19）
道：天工开物
理：几何次序、数字阵列
法：梭织二方连缀、纬向接回5厘米
术：针织组织、梭织绉纱组织、轧花、切割
势：2014/2015国际流行趋势之民族文化形态
器：服装面料、装饰布

Name: Golden Diamond-shape Spearhead (17), (18), (19)
Conviction: heavenly creations
Principle: geometric sequence, digital array
Method: weaving 2 sides in continuation, zonal tying-in 5 cm
Operation: knitting, weaving crepe texture, embossing, cutting
Trend: 2014/2015 international fashion trend of national culture
Application: clothing fabrics, decorative cloth

时间：2010年11月29日
地点：上海淮海东路卡地亚专卖店
对象：幕墙
灵感：阵列、光感、色感
Time: November 29, 2010
Location: Cartire stores of East Huaihai Road, Shanghai
Object: curtain wall
Inspiration: array, light, color

Creative Fabrics Design 1

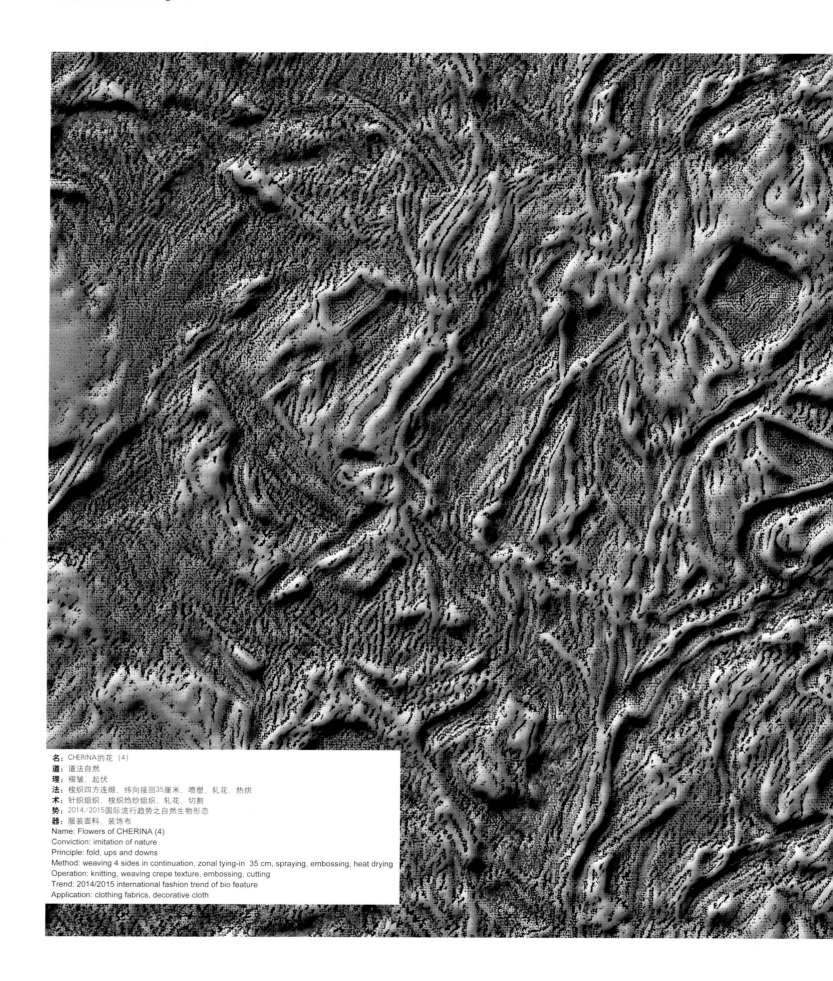

名：CHERINA的花（4）
道：道法自然
理：褶皱、起伏
法：梭织四方连缀、纬向接回35厘米、喷塑、轧花、热烘
术：针织组织，梭织绉纱组织、轧花、切割
势：2014/2015国际流行趋势之自然生物形态
器：服装面料、装饰布

Name: Flowers of CHERINA (4)
Conviction: imitation of nature
Principle: fold, ups and downs
Method: weaving 4 sides in continuation, zonal tying-in 35 cm, spraying, embossing, heat drying
Operation: knitting, weaving crepe texture, embossing, cutting
Trend: 2014/2015 international fashion trend of bio feature
Application: clothing fabrics, decorative cloth

时间：2010年10月29日
地点：尼泊尔奇旺国家野生动物园
对象：大象皮肤
灵感：阳光下皱褶的质感
Time: October 29, 2010
Location: The Chitwan National Park, Nepal
Object: the skin of elephant
Inspiration: wrinkle texture under the sunlight

Creative Fabrics Design 1

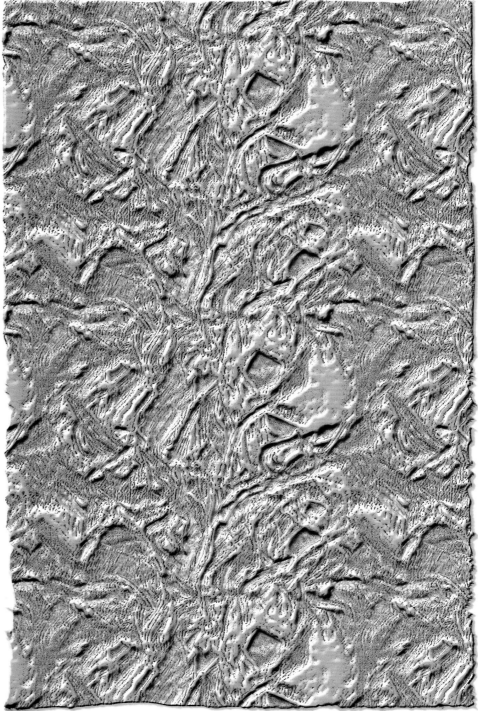
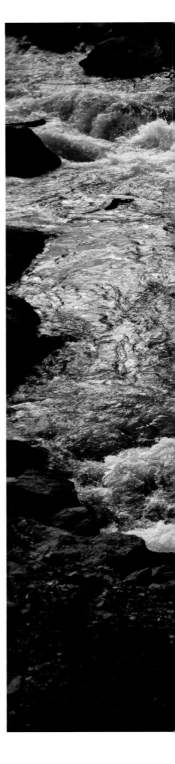

名: CHERINA的花 (5)
道: 道法自然
理: 褶皱、起伏、银光
法: 梭织四方连缀、纬向接回35厘米、喷塑、轧花、热烘
术: 针织组织、梭织绉纱组织、轧花、切割
势: 2014/2015国际流行趋势之自然生物形态
器: 服装面料、装饰布

Name: Flowers of CHERINA (4)
Conviction: imitation of nature
Principle: fold, ups and downs
Method: weaving 4 sides in continuation, zonal tying-in 35 cm, spray, embossing, heat drying
Operation: knitting, weaving crepe texture, embossing, cutting
Trend: 2014/2015 international fashion trend of bio feature
Application: clothing fabrics, decorative cloth

时间：2007年7月29日
地点：四川川藏公路
对象：湍急的河流
灵感：波光粼粼、奔腾之感
Time: July 29, 2007
Location: Sichuan-Tibetan Highway, Sichuan
Object: the rapids of the river
Inspiration: glittering surges

Creative Fabrics Design 1

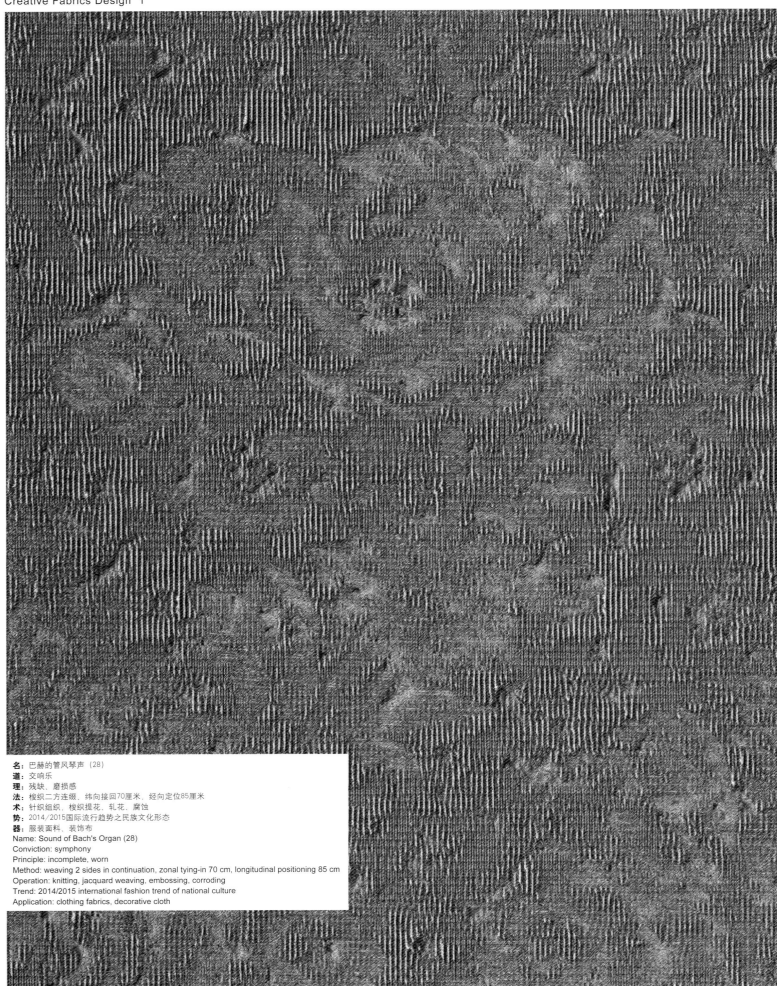

名：巴赫的管风琴声（28）
道：交响乐
理：残缺、磨损感
法：梭织二方连缀、纬向接回70厘米、经向定位85厘米
术：针织组织，梭织提花、轧花、腐蚀
势：2014/2015国际流行趋势之民族文化形态
器：服装面料、装饰布

Name: Sound of Bach's Organ (28)
Conviction: symphony
Principle: incomplete, worn
Method: weaving 2 sides in continuation, zonal tying-in 70 cm, longitudinal positioning 85 cm
Operation: knitting, jacquard weaving, embossing, corroding
Trend: 2014/2015 international fashion trend of national culture
Application: clothing fabrics, decorative cloth

时间：2010年10月25日
地点：尼泊尔奇望巴德岗
对象：木雕门
灵感：风化

Time: October 25, 2010
Location: Bhadgaon, Nepal
Objects: engraved wooden door
Inspiration: weathering

Creative Fabrics Design 1

名：巴赫的管风琴声 (29)
道：交响乐
理：残缺、磨损感
法：梭织二方连缀、纬向接回70厘米，经向定位85厘米
术：针织组织，梭织提花、轧花、腐蚀
势：2014/2015国际流行趋势之民族文化形态
器：服装面料、装饰布

Name: Sound of Bach's Organ (29)
Conviction: symphony
Principle: incomplete, worn
Method: weaving 2 sides in continuation, zonal tying-in tying 70 cm, longitudinal positioning 85 cm
Operation: knitting, jacquard weaving, embossing, corroding
Trend: 2014/2015 international fashion trend of national culture
Application: clothing fabrics, decorative cloth

时间：2010年10月25日
地点：尼泊尔奇望巴德岗
对象：木雕门
灵感：风化
Time: October 25, 2010
Location: Bhadgaon, Nepal
Objects: engraved wooden door
Inspiration: weathering

Creative Fabrics Design 1

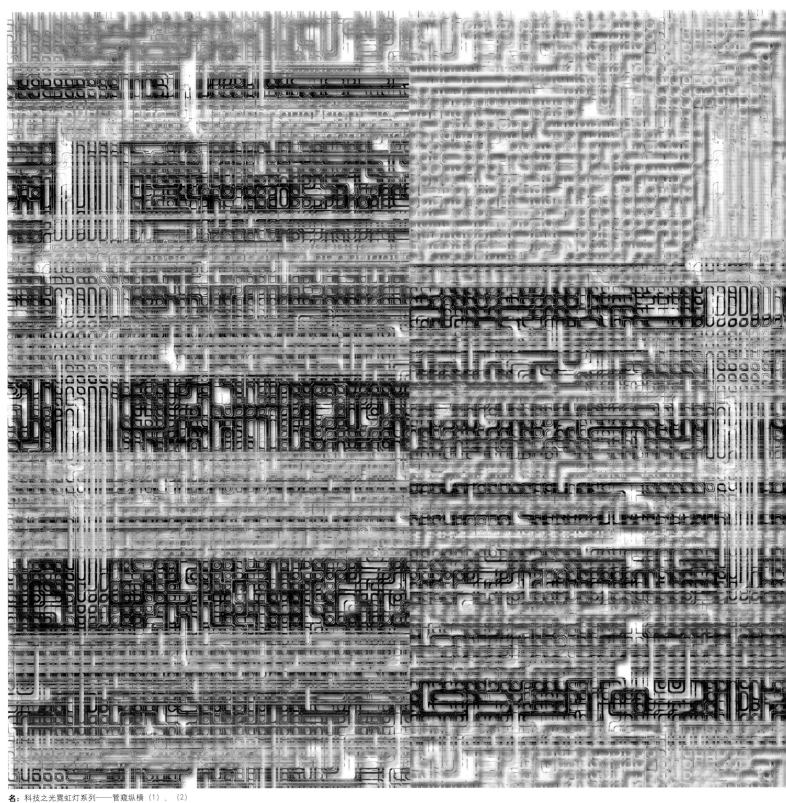

名：科技之光霓虹灯系列——管窥纵横（1）、（2）
道：天工开物
理：几何次序，数字阵列
法：印花，梭织二方连缀，纬向接回21.6厘米、经向定位85厘米
术：电子线路，荧光线路，梭织提花反面浮线组织
势：2014/2015国际流行趋势之科技形态
器：服装面料，装饰布

Name: Light of Science and Technology — Neon Lamp Series （1）,（2）
Conviction: heavenly creations
Principle: geometric sequence, digital array
Method: printing, weaving 2 sides in continuation, zonal tying-in 21.6 cm, longitudinal tying-in 85 cm
Operation: electronic circuit, fluorescence line, weaving jacquard of loose stitch on back side
Trend: 2014/2015 international fashion trend in science and technology
Application: clothing fabrics, decorative cloth

时间：2010年11月29日
地点：上海淮海东路
对象：办公楼灯幕墙
灵感：阵列、光感
Time: November 29, 2010
Location: East Huaihai Road, Shanghai
Object: curtain wall of office building
Inspiration: array, light

Creative Fabrics Design 1

名：花皇（2）
道：天工开物、道法自然
理：木刻、浮雕
法：梭织二方连缀、纬向接回35厘米，经向定位85厘米
术：梭织提花、印染、轧花、腐蚀
势：2014/2015国际流行趋势之民族文化形态
器：服装面料、装饰布
Name: the Queen of Flowers (2)
Conviction: heavenly creations, imitation of nature
Principle: wood carving, relief
Method: weaving 2 sides in continuation, zonal tying-in 35 cm, longitudinal positioning 85 cm
Operation: jacquard weaving, printing and dyeing, embossing, corroding
Trend: 2014/2015 international fashion trend of national culture
Application: clothing fabrics, decorative cloth

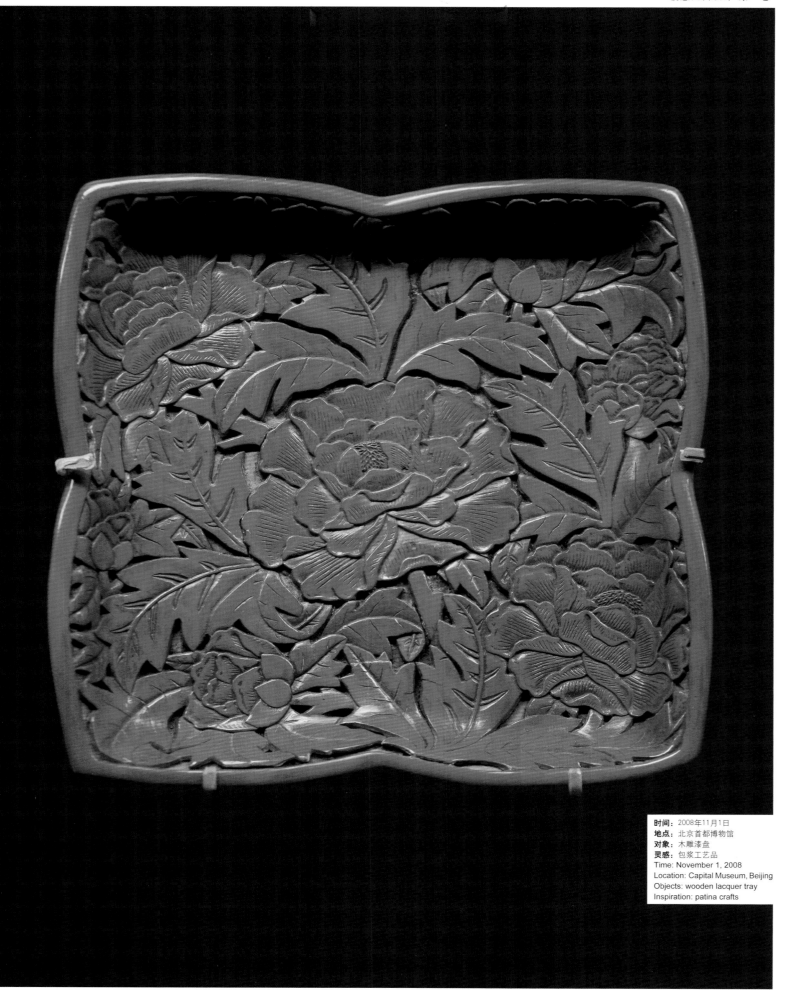

时间：2008年11月1日
地点：北京首都博物馆
对象：木雕漆盘
灵感：包浆工艺品
Time: November 1, 2008
Location: Capital Museum, Beijing
Objects: wooden lacquer tray
Inspiration: patina crafts

Creative Fabrics Design 1

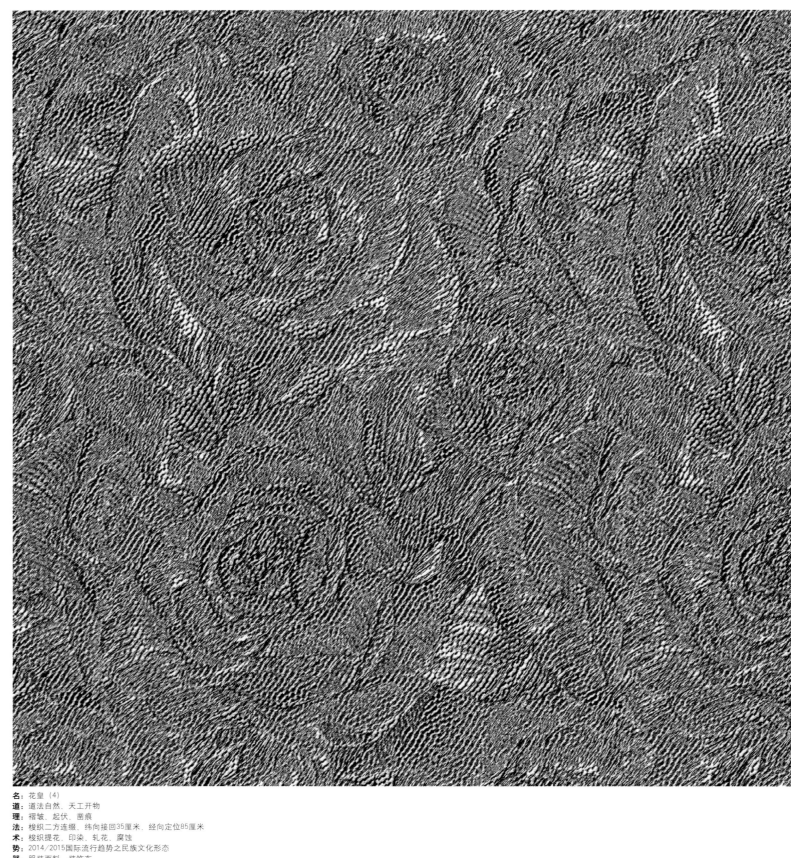

名：花皇（4）
道：道法自然、天工开物
理：褶皱、起伏、凿痕
法：梭织二方连缀、纬向接回35厘米、经向定位85厘米
术：梭织提花、印染、轧花、腐蚀
势：2014/2015国际流行趋势之民族文化形态
器：服装面料、装饰布

Name: the Queen of Flowers (4)
Conviction: nature, heavenly creations
Principle: fold, ups and downs, chisel marks
Method: weaving 2 sides in continuation, zonal tying-in 35 cm, longitudinal positioning 85 cm
Operation: jacquard weaving, printing and dyeing, embossing, corroding
Trend: 2014/2015 international fashion trend of national culture
Application: clothing fabrics, decorative cloth

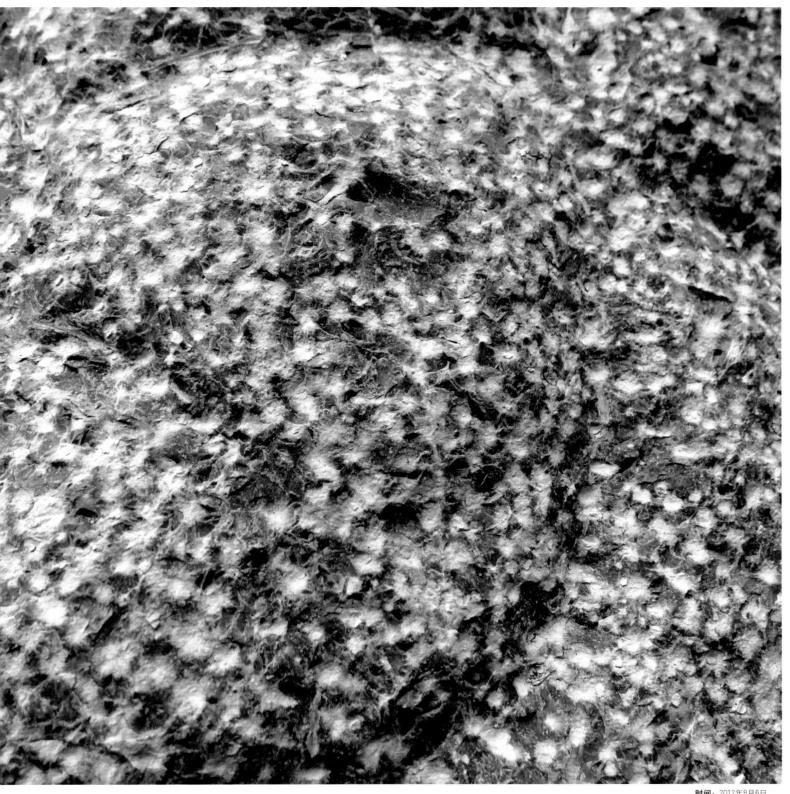

时间：2012年8月6日
地点：云南丽江
对象：石雕细部
灵感：凿斑
Time: August 6, 2012
Location: Lijiang, Yunnan
Object: detail of stone
Inspiration: chisel spot

Creative Fabrics Design 1

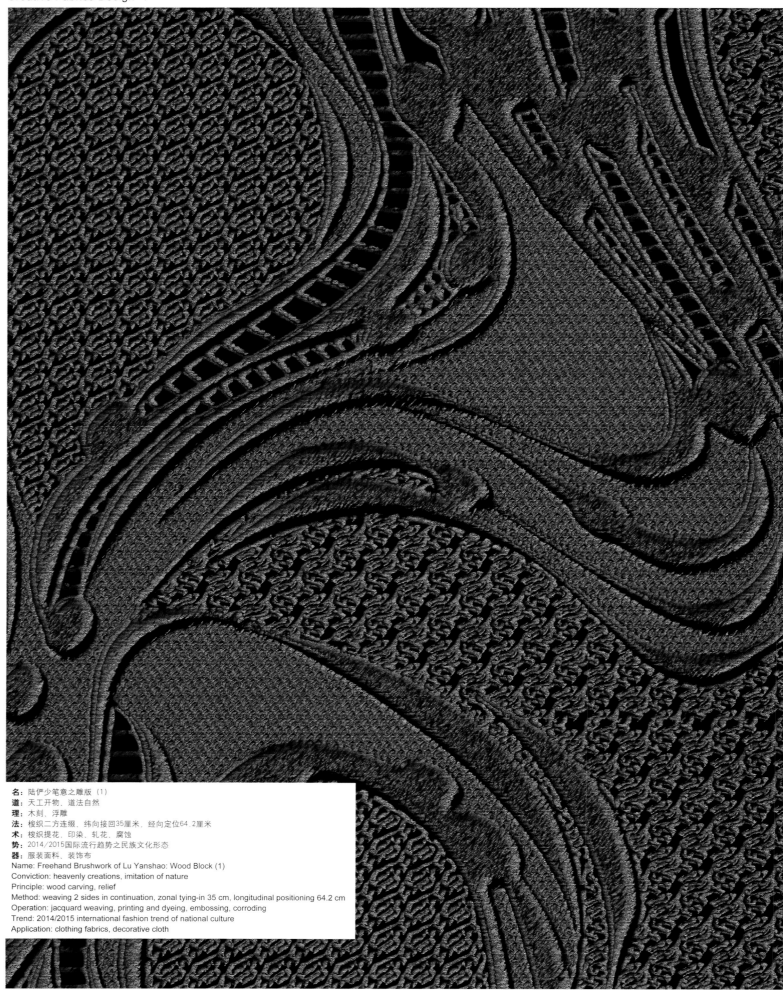

名：陆俨少笔意之雕版（1）
道：天工开物，道法自然
理：木刻，浮雕
法：梭织二方连缀，纬向接回35厘米，经向定位64.2厘米
术：梭织提花，印染，轧花，腐蚀
势：2014/2015国际流行趋势之民族文化形态
器：服装面料，装饰布

Name: Freehand Brushwork of Lu Yanshao: Wood Block (1)
Conviction: heavenly creations, imitation of nature
Principle: wood carving, relief
Method: weaving 2 sides in continuation, zonal tying-in 35 cm, longitudinal positioning 64.2 cm
Operation: jacquard weaving, printing and dyeing, embossing, corroding
Trend: 2014/2015 international fashion trend of national culture
Application: clothing fabrics, decorative cloth

时间：2009年8月3日
地点：北京首都博物馆
对象：木雕漆盘
灵感：包浆工艺品
Time: August 3, 2009
Location: Capital Museum, Beijing
Objects: wooded lacquer tray
Inspiration: patina crafts

Creative Fabrics Design 1

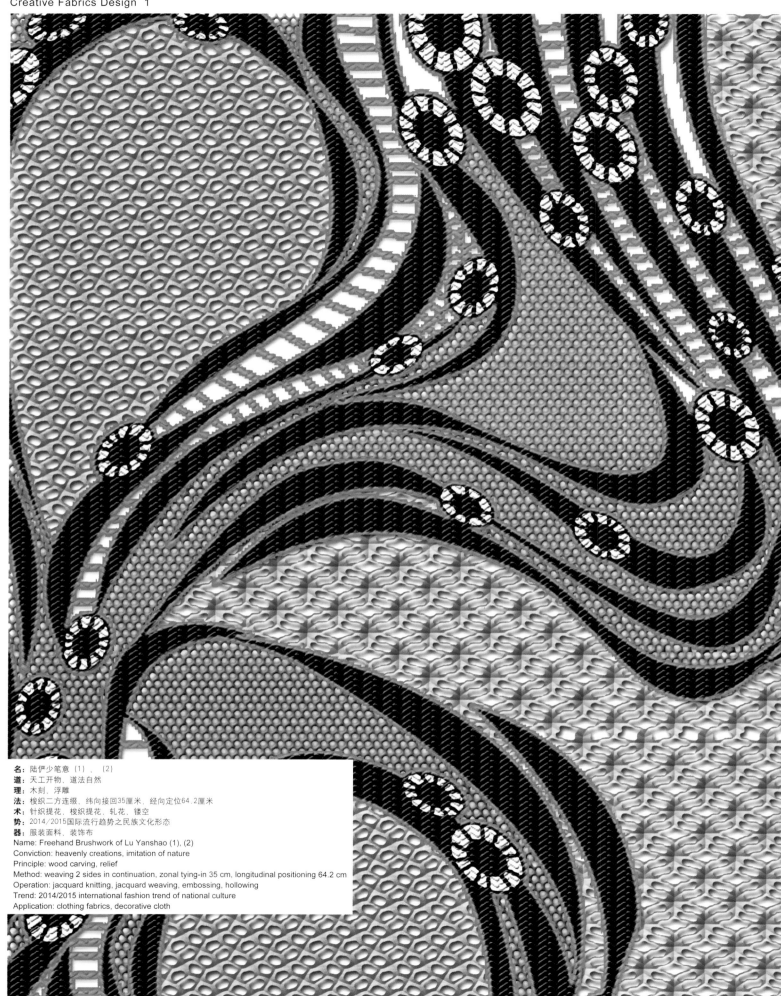

名：陆俨少笔意（1）、（2）
道：天工开物，道法自然
理：木刻，浮雕
法：梭织二方连缀、纬向接回35厘米，经向定位64.2厘米
术：针织提花，梭织提花，轧花，镂空
势：2014/2015国际流行趋势之民族文化形态
器：服装面料、装饰布
Name: Freehand Brushwork of Lu Yanshao (1), (2)
Conviction: heavenly creations, imitation of nature
Principle: wood carving, relief
Method: weaving 2 sides in continuation, zonal tying-in 35 cm, longitudinal positioning 64.2 cm
Operation: jacquard knitting, jacquard weaving, embossing, hollowing
Trend: 2014/2015 international fashion trend of national culture
Application: clothing fabrics, decorative cloth

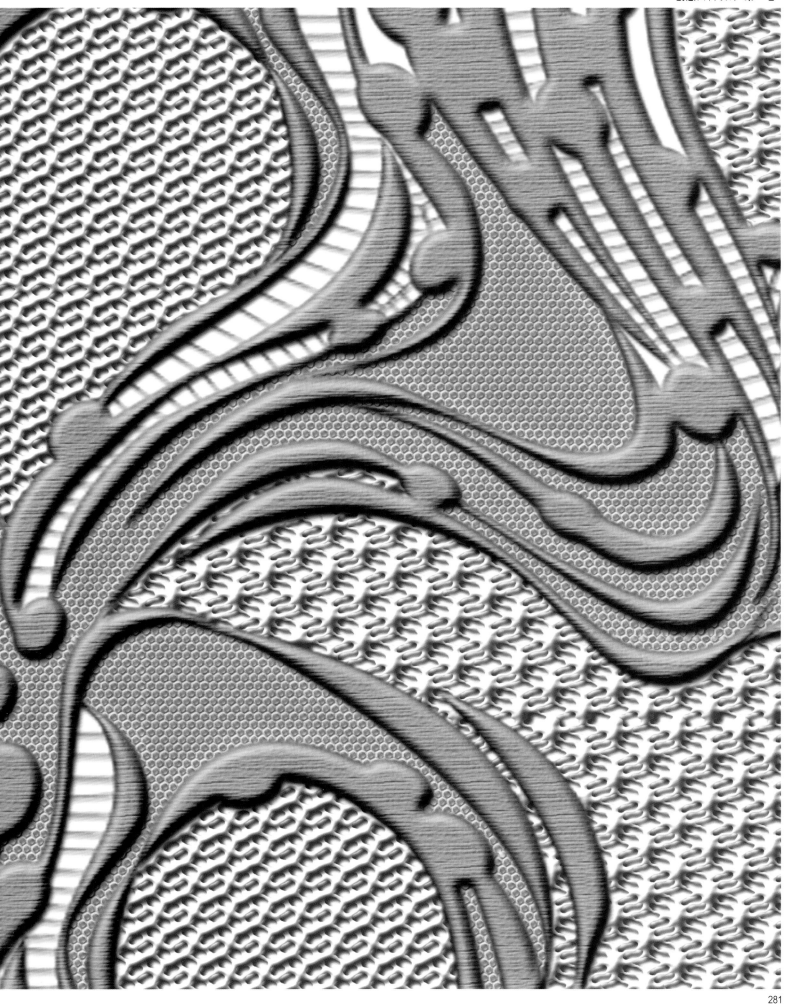

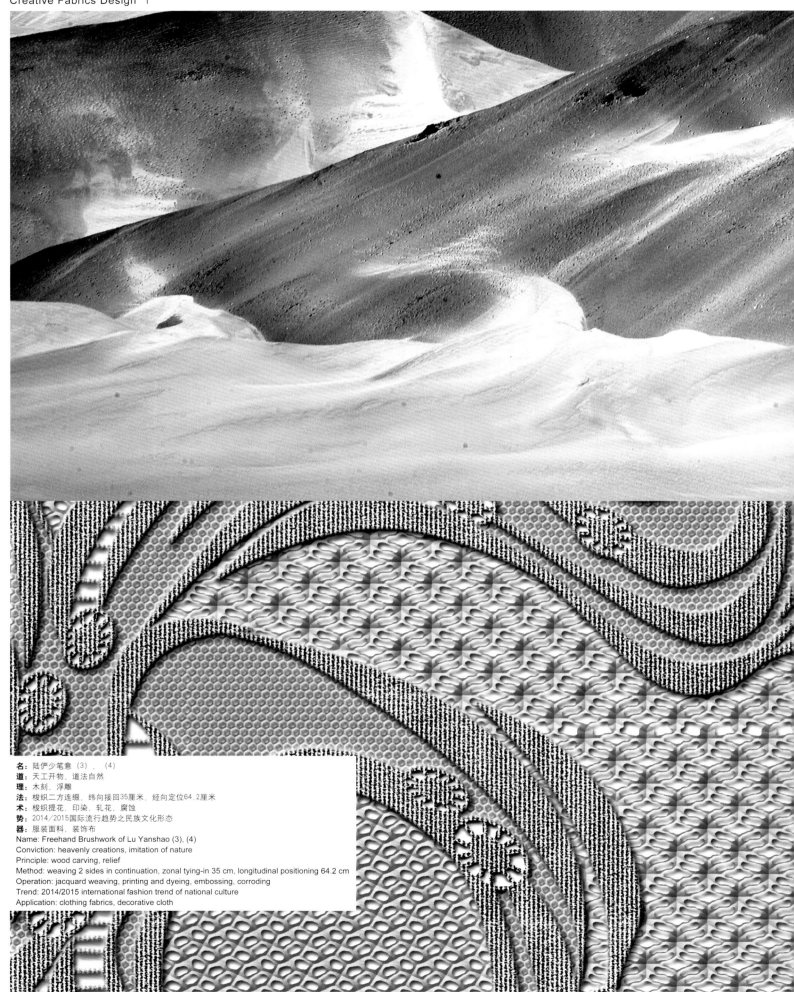

名：陆俨少笔意（3）、（4）
道：天工开物，道法自然
理：木刻，浮雕
法：梭织二方连缀，纬向接回35厘米，经向定位64.2厘米
术：梭织提花，印染，轧花，腐蚀
势：2014/2015国际流行趋势之民族文化形态
器：服装面料，装饰布

Name: Freehand Brushwork of Lu Yanshao (3), (4)
Conviction: heavenly creations, imitation of nature
Principle: wood carving, relief
Method: weaving 2 sides in continuation, zonal tying-in 35 cm, longitudinal positioning 64.2 cm
Operation: jacquard weaving, printing and dyeing, embossing, corroding
Trend: 2014/2015 international fashion trend of national culture
Application: clothing fabrics, decorative cloth

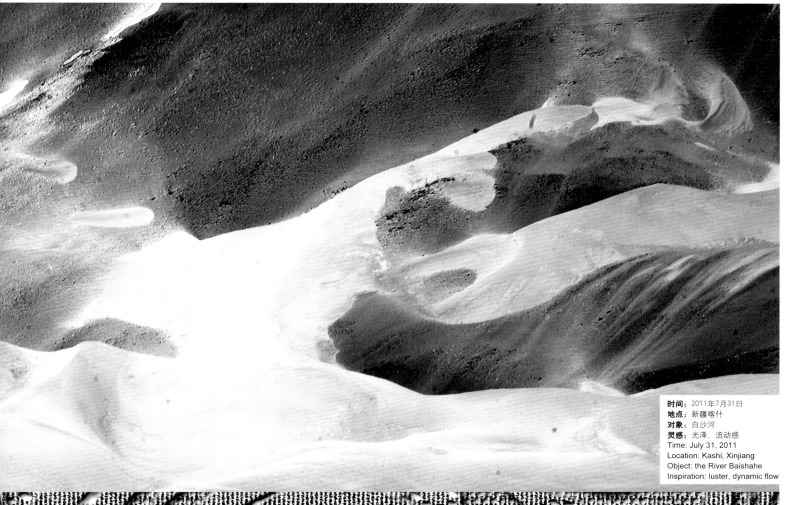

时间: 2011年7月31日
地点: 新疆喀什
对象: 白沙河
灵感: 光泽，流动感
Time: July 31, 2011
Location: Kashi, Xinjiang
Object: the River Baishahe
Inspiration: luster, dynamic flow

Creative Fabrics Design 1

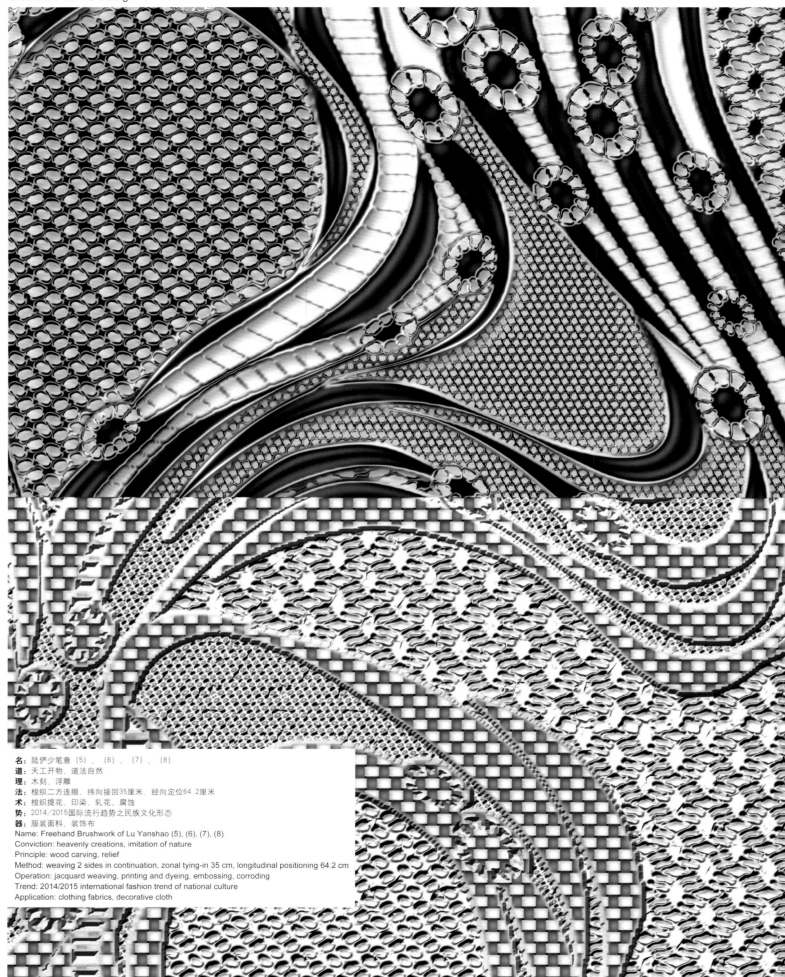

名：陆俨少笔意（5）、（6）、（7）、（8）
道：天工开物，道法自然
理：木刻，浮雕
法：梭织二方连缀，纬向接回35厘米，经向定位64.2厘米
术：梭织提花、印染、轧花、腐蚀
势：2014/2015国际流行趋势之民族文化形态
器：服装面料、装饰布

Name: Freehand Brushwork of Lu Yanshao (5), (6), (7), (8)
Conviction: heavenly creations, imitation of nature
Principle: wood carving, relief
Method: weaving 2 sides in continuation, zonal tying-in 35 cm, longitudinal positioning 64.2 cm
Operation: jacquard weaving, printing and dyeing, embossing, corroding
Trend: 2014/2015 international fashion trend of national culture
Application: clothing fabrics, decorative cloth

Creative Fabrics Design 1

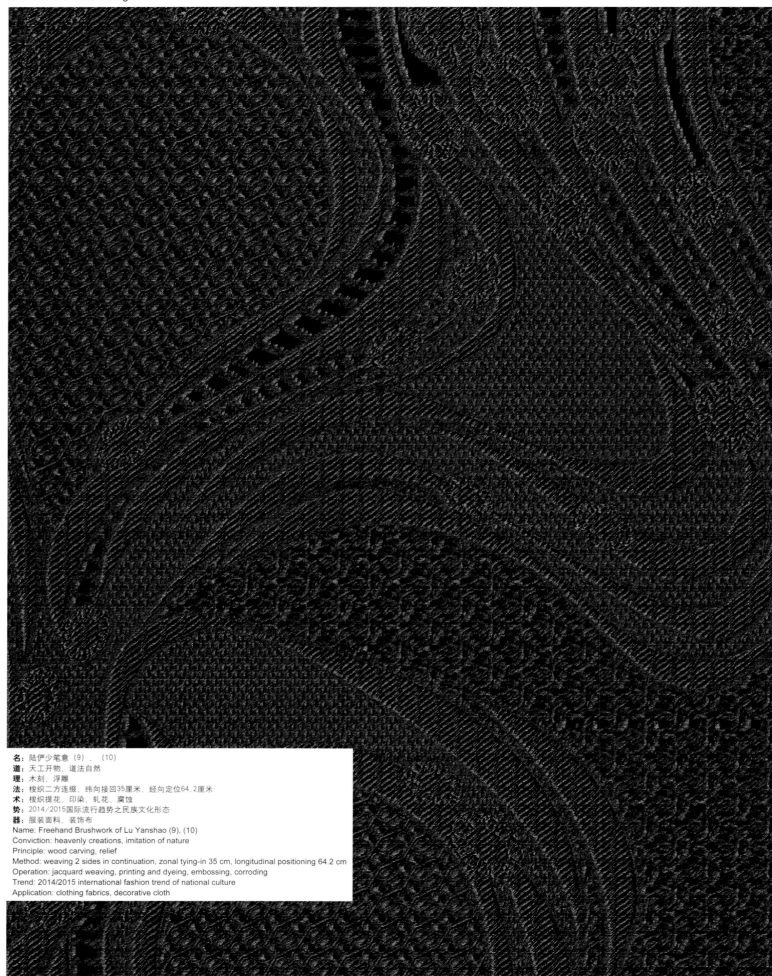

名：陆俨少笔意（9）、（10）
道：天工开物、道法自然
理：木刻、浮雕
法：梭织二方连缀，纬向接回35厘米，经向定位64.2厘米
术：梭织提花、印染、轧花、腐蚀
势：2014/2015国际流行趋势之民族文化形态
器：服装面料、装饰布

Name: Freehand Brushwork of Lu Yanshao (9), (10)
Conviction: heavenly creations, imitation of nature
Principle: wood carving, relief
Method: weaving 2 sides in continuation, zonal tying-in 35 cm, longitudinal positioning 64.2 cm
Operation: jacquard weaving, printing and dyeing, embossing, corroding
Trend: 2014/2015 international fashion trend of national culture
Application: clothing fabrics, decorative cloth

Creative Fabrics Design 1

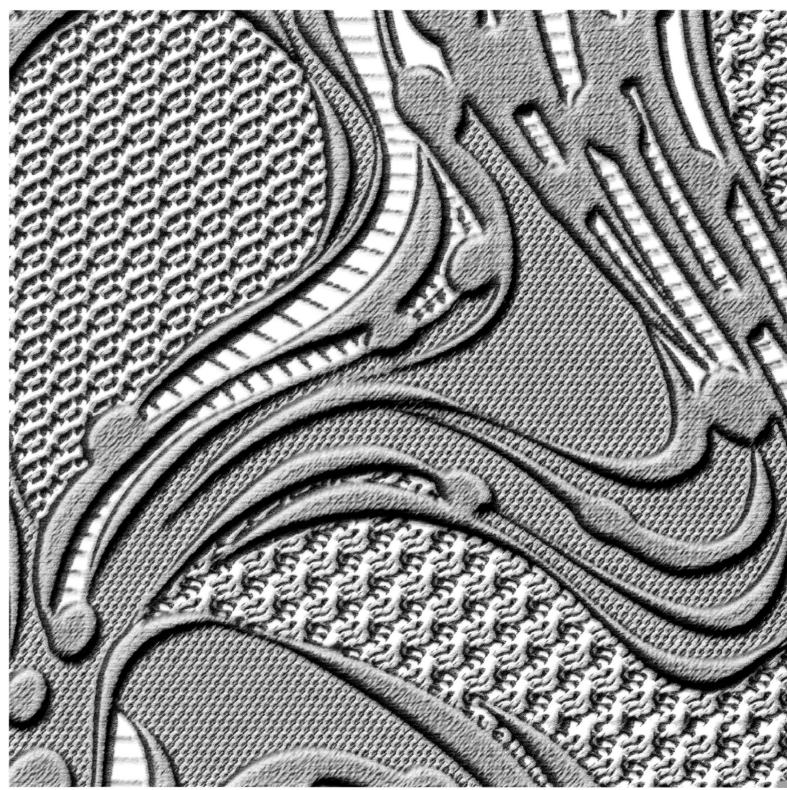

名：陆俨少笔意（11）
道：天工开物，道法自然
理：木刻、浮雕
法：梭织二方连缀，纬向接回35厘米，经向定位64.2厘米
术：梭织提花、印染、轧花、腐蚀
势：2014/2015国际流行趋势之民族文化形态
器：服装面料、装饰布

Name: Freehand Brushwork of Lu Yanshao (11)
Conviction: heavenly creations, imitation of nature
Principle: wood carving, relief
Method: weaving 2 sides in continuation, zonal tying-in 35 cm, longitudinal positioning 64.2 cm
Operation: jacquard weaving, printing and dyeing, embossing, corroding
Trend: 2014/2015 international fashion trend of national culture
Application: clothing fabrics, decorative cloth

创意面料设计 第一卷

时间：2009年7月19日
地点：青海茶卡
对象：盐湖
灵感：矿物质
Time: July 19, 2009
Location: Chaka,Qinghai
Object: salt lake
Inspiration: mineral

Creative Fabrics Design 1

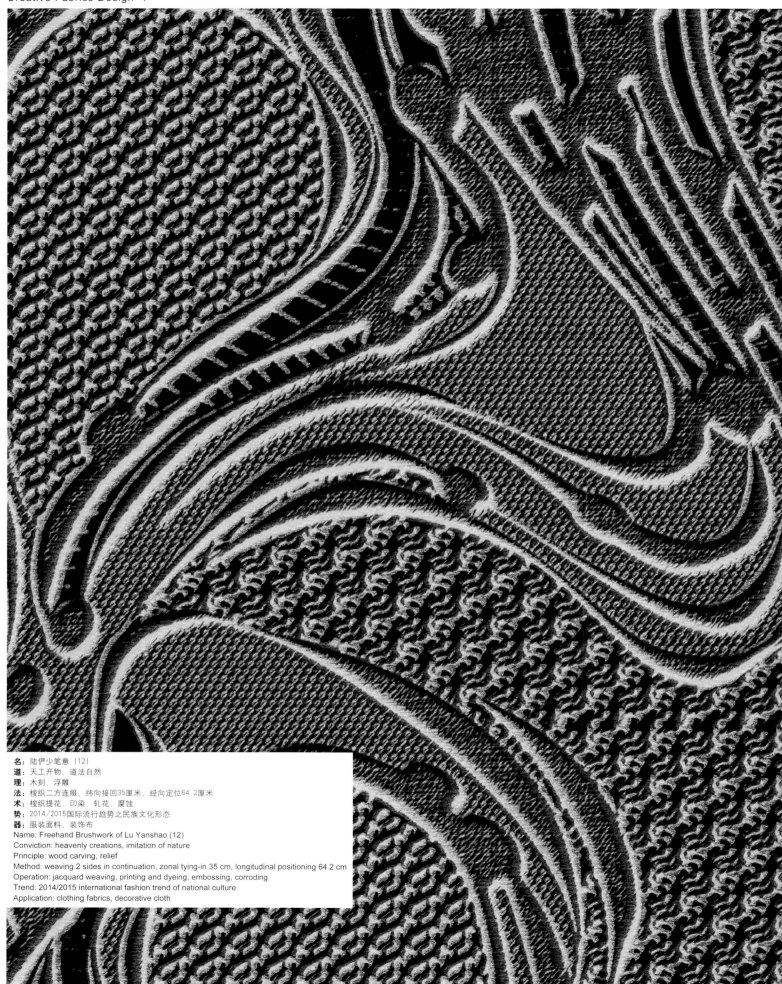

名：陆俨少笔意（12）
道：天工开物，道法自然
理：木刻、浮雕
法：梭织二方连缀，纬向接回35厘米，经向定位64.2厘米
术：梭织提花、印染、轧花、腐蚀
势：2014/2015国际流行趋势之民族文化形态
器：服装面料、装饰布

Name: Freehand Brushwork of Lu Yanshao (12)
Conviction: heavenly creations, imitation of nature
Principle: wood carving, relief
Method: weaving 2 sides in continuation, zonal tying-in 35 cm, longitudinal positioning 64.2 cm
Operation: jacquard weaving, printing and dyeing, embossing, corroding
Trend: 2014/2015 international fashion trend of national culture
Application: clothing fabrics, decorative cloth

时间：2012年8月2日
地点：四川宜宾
对象：浮雕纪念碑
灵感：肃然与庄重
Time: August 2, 2012
Location: Yibin, Sichuan
Object: relief monument
Inspiration: respectfulness and dignity

Creative Fabrics Design 1

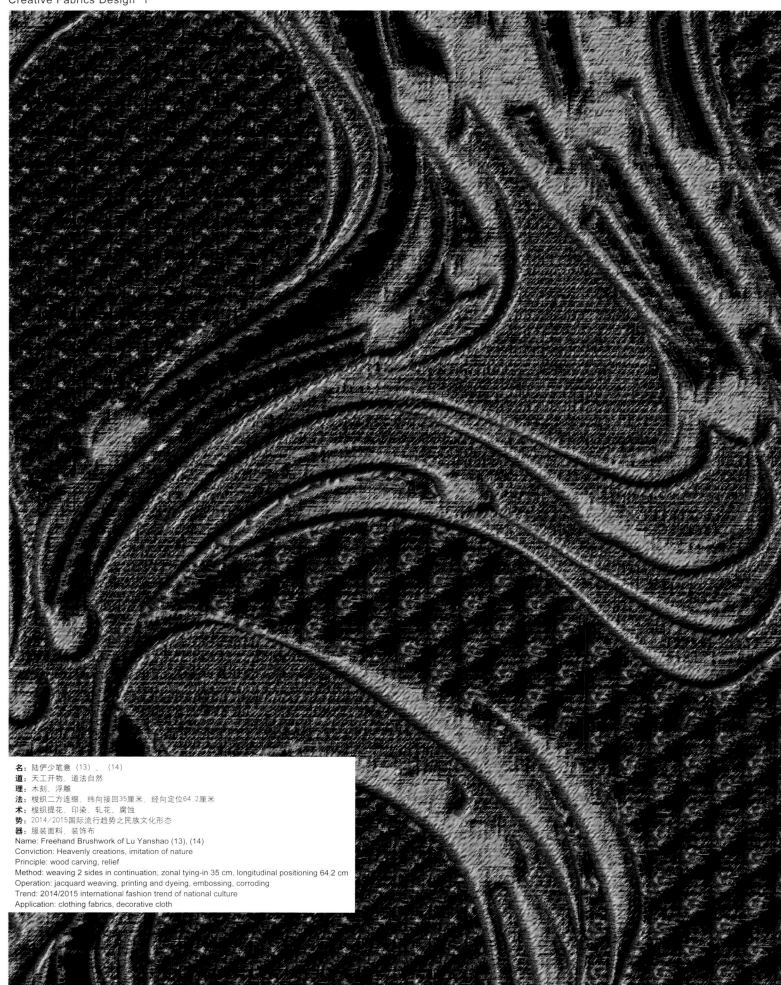

名：陆俨少笔意（13）、（14）
道：天工开物，道法自然
理：木刻、浮雕
法：梭织二方连缀，纬向接回35厘米，经向定位64.2厘米
术：梭织提花、印染、轧花、腐蚀
势：2014/2015国际流行趋势之民族文化形态
器：服装面料、装饰布
Name: Freehand Brushwork of Lu Yanshao (13), (14)
Conviction: Heavenly creations, imitation of nature
Principle: wood carving, relief
Method: weaving 2 sides in continuation, zonal tying-in 35 cm, longitudinal positioning 64.2 cm
Operation: jacquard weaving, printing and dyeing, embossing, corroding
Trend: 2014/2015 international fashion trend of national culture
Application: clothing fabrics, decorative cloth

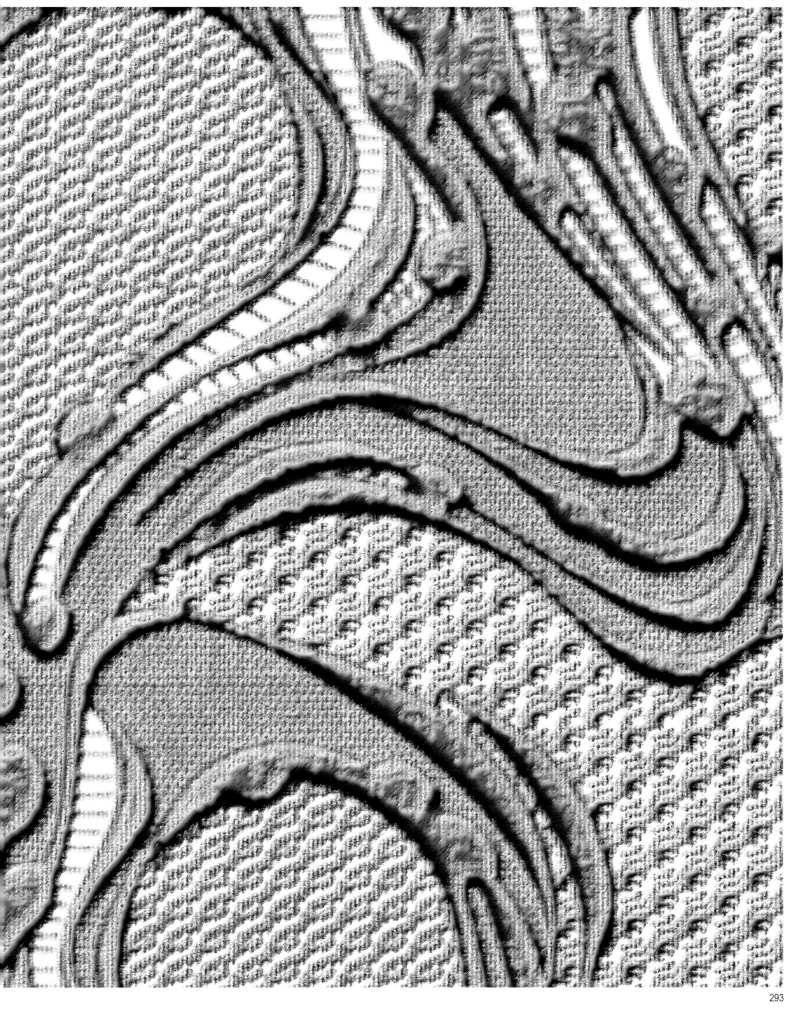

Creative Fabrics Design 1

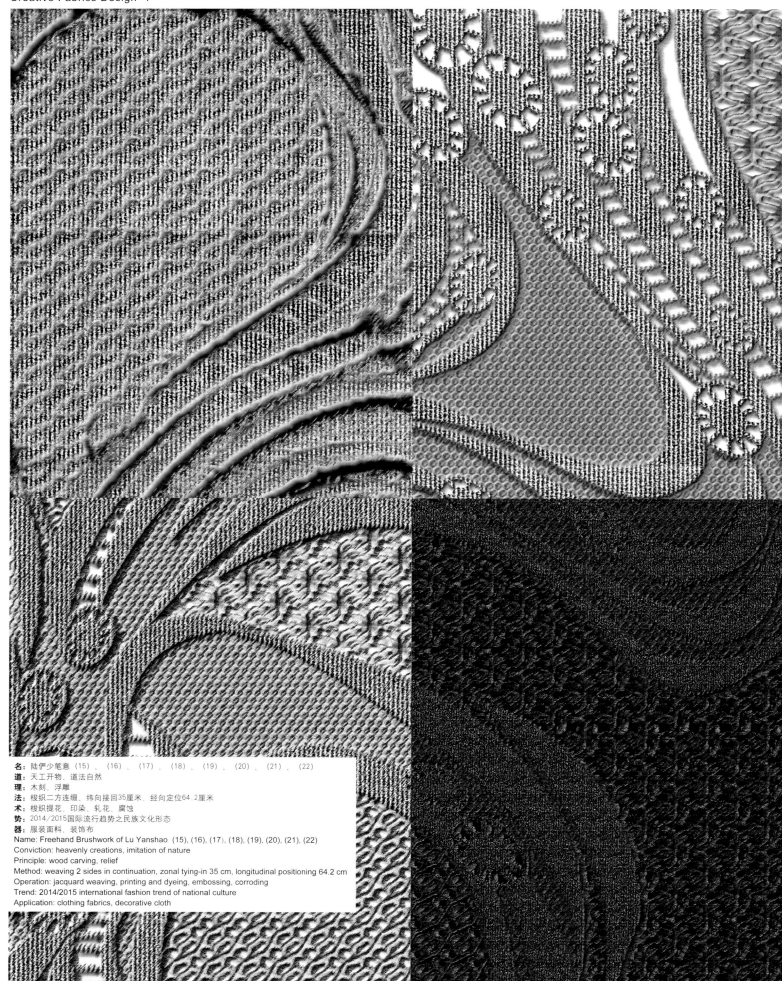

名：陆俨少笔意（15）、（16）、（17）、（18）、（19）、（20）、（21）、（22）
道：天工开物、道法自然
理：木刻、浮雕
法：梭织二方连缀、纬向接回35厘米、经向定位64.2厘米
术：梭织提花、印染、轧花、腐蚀
势：2014/2015国际流行趋势之民族文化形态
器：服装面料、装饰布

Name: Freehand Brushwork of Lu Yanshao (15), (16), (17), (18), (19), (20), (21), (22)
Conviction: heavenly creations, imitation of nature
Principle: wood carving, relief
Method: weaving 2 sides in continuation, zonal tying-in 35 cm, longitudinal positioning 64.2 cm
Operation: jacquard weaving, printing and dyeing, embossing, corroding
Trend: 2014/2015 international fashion trend of national culture
Application: clothing fabrics, decorative cloth

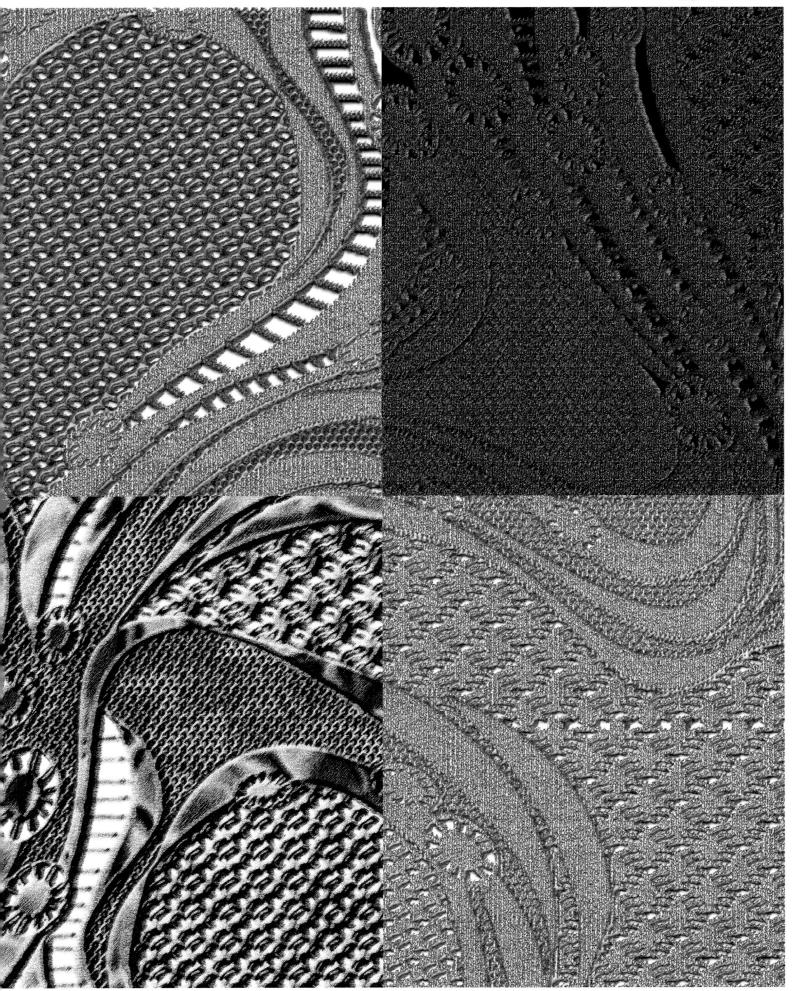

Creative Fabrics Design 1

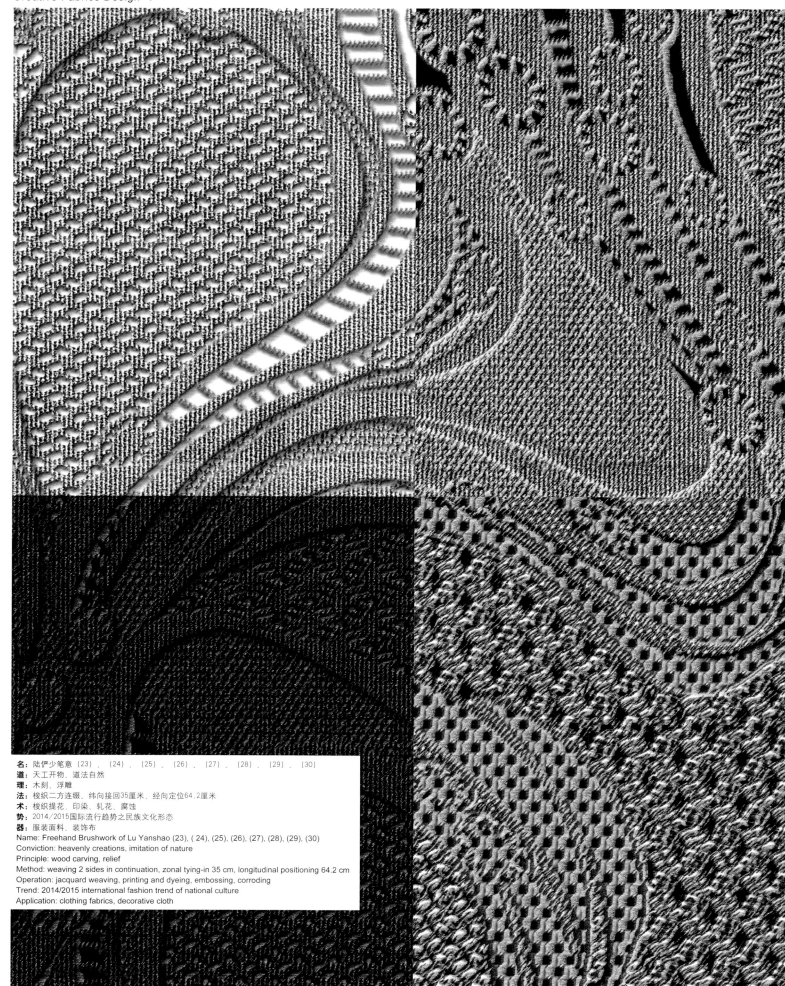

名：陆俨少笔意（23）、（24）、（25）、（26）、（27）、（28）、（29）、（30）
道：天工开物、道法自然
理：木刻、浮雕
法：梭织二方连缀、纬向接回35厘米，经向定位64.2厘米
术：梭织提花、印染、轧花、腐蚀
势：2014/2015国际流行趋势之民族文化形态
器：服装面料、装饰布

Name: Freehand Brushwork of Lu Yanshao (23), (24), (25), (26), (27), (28), (29), (30)
Conviction: heavenly creations, imitation of nature
Principle: wood carving, relief
Method: weaving 2 sides in continuation, zonal tying-in 35 cm, longitudinal positioning 64.2 cm
Operation: jacquard weaving, printing and dyeing, embossing, corroding
Trend: 2014/2015 international fashion trend of national culture
Application: clothing fabrics, decorative cloth

Creative Fabrics Design 1

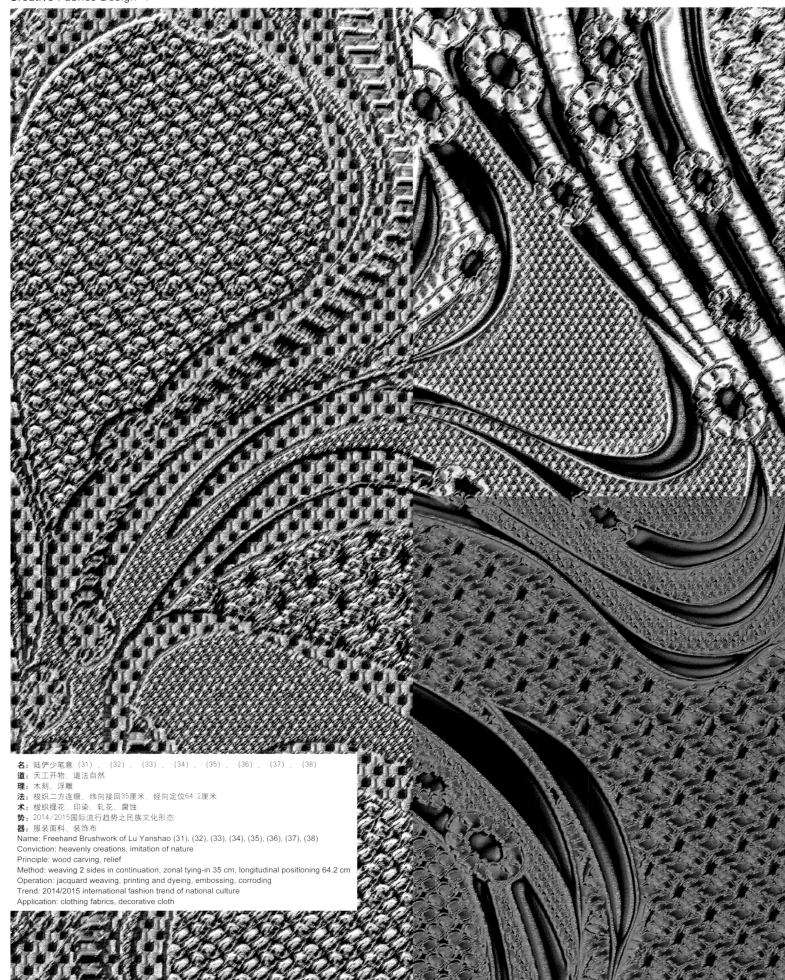

名：陆俨少笔意（31）、（32）、（33）、（34）、（35）、（36）、（37）、（38）
道：天工开物、道法自然
理：木刻、浮雕
法：梭织二方连缀，纬向接回35厘米，经向定位64.2厘米
术：梭织提花、印染、轧花、腐蚀
势：2014/2015国际流行趋势之民族文化形态
器：服装面料、装饰布

Name: Freehand Brushwork of Lu Yanshao (31), (32), (33), (34), (35), (36), (37), (38)
Conviction: heavenly creations, imitation of nature
Principle: wood carving, relief
Method: weaving 2 sides in continuation, zonal tying-in 35 cm, longitudinal positioning 64.2 cm
Operation: jacquard weaving, printing and dyeing, embossing, corroding
Trend: 2014/2015 international fashion trend of national culture
Application: clothing fabrics, decorative cloth

Creative Fabrics Design 1

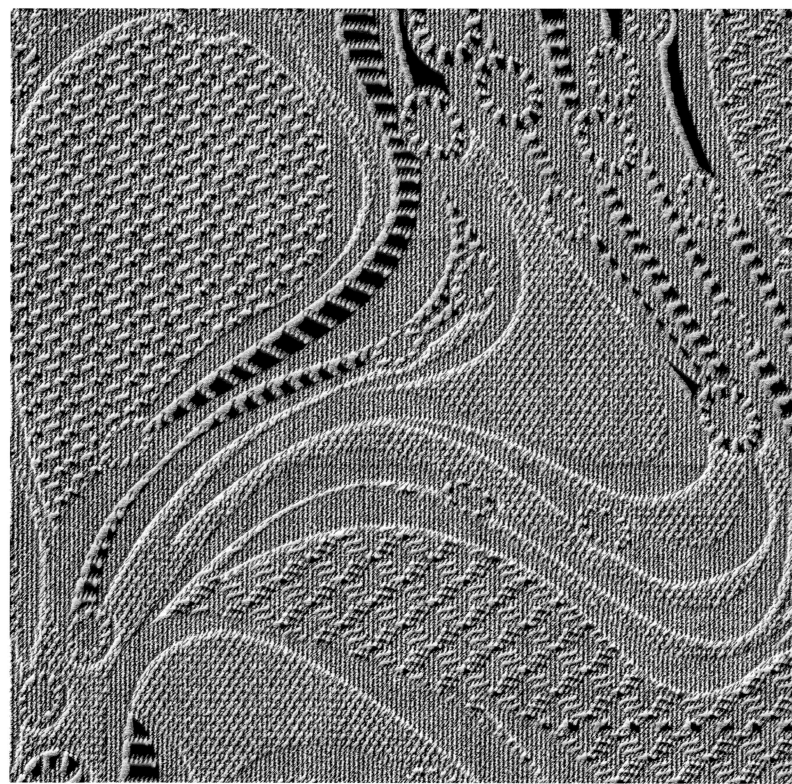

名: 陆俨少笔意 (39)
道: 天工开物、道法自然
理: 石刻、浮雕
法: 梭织二方连缀、纬向接回35厘米,经向定位64.2厘米
术: 梭织提花、印染、轧花、腐蚀
势: 2014/2015国际流行趋势之民族文化形态
器: 服装面料、装饰布

Name: Freehand Brushwork of Lu Yanshao (39)
Conviction: heavenly creations, imitation of nature
Principle: wood carving, relief
Method: weaving 2 sides in continuation, zonal tying-in 35 cm, longitudinal positioning 64.2 cm
Operation: jacquard weaving, printing and dyeing, embossing, corroding
Trend: 2014/2015 international fashion trend of national culture
Application: clothing fabrics, decorative cloth

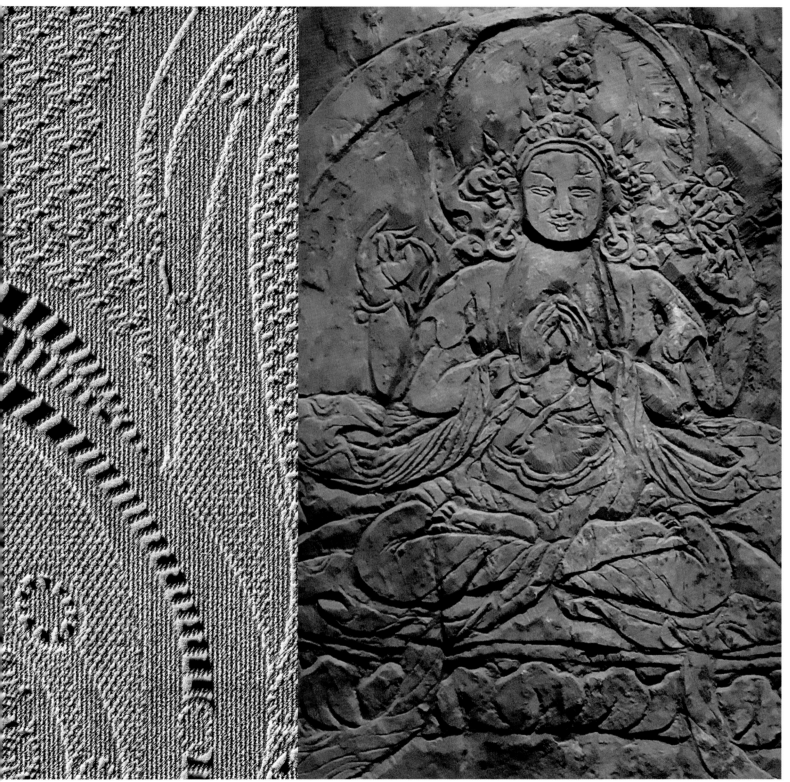

时间： 2012年8月8日
地点： 云南中甸
对象： 玛尼石刻
灵感： 凿斑

Time: August 8, 2012
Location: Zhongdian, Yunnan
Object: Mani stone carvings
Inspiration: chisel spot

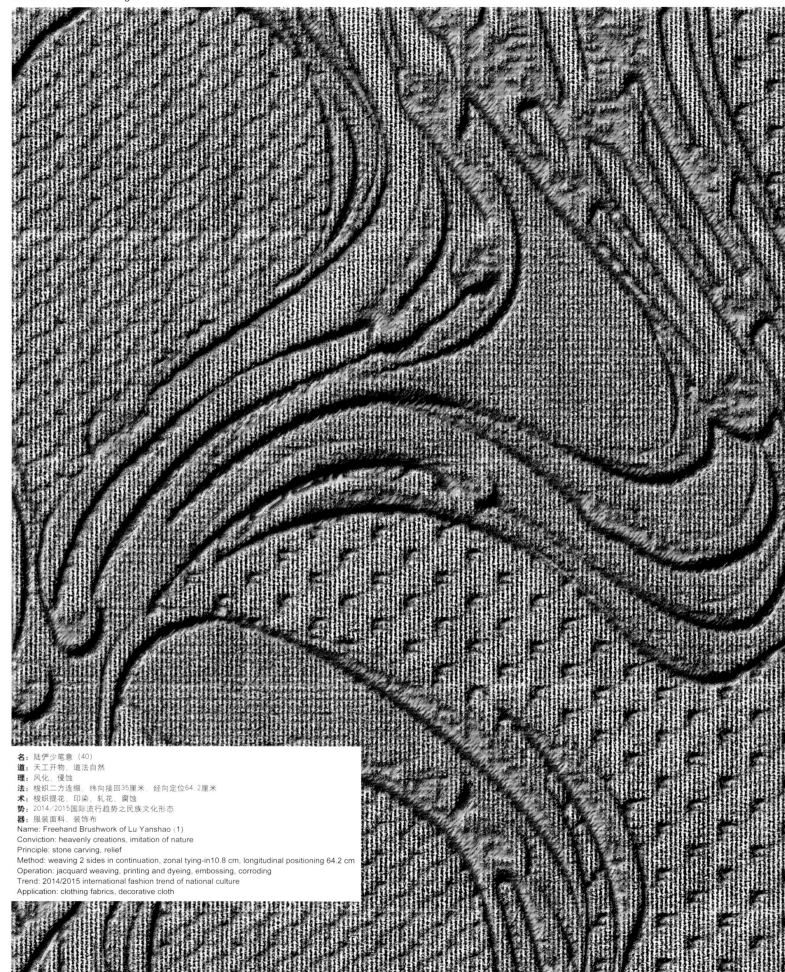

名：陆俨少笔意（40）
道：天工开物，道法自然
理：风化，侵蚀
法：梭织二方连缀，纬向接回35厘米，经向定位64.2厘米
术：梭织提花，印染、轧花、腐蚀
势：2014/2015国际流行趋势之民族文化形态
器：服装面料、装饰布

Name: Freehand Brushwork of Lu Yanshao (1)
Conviction: heavenly creations, imitation of nature
Principle: stone carving, relief
Method: weaving 2 sides in continuation, zonal tying-in 10.8 cm, longitudinal positioning 64.2 cm
Operation: jacquard weaving, printing and dyeing, embossing, corroding
Trend: 2014/2015 international fashion trend of national culture
Application: clothing fabrics, decorative cloth

时间：2012年7月31日
地点：四川成都宽窄巷子
对象：青砖墙
灵感：经年累月
Time: July 31, 2012
Location: Broad and Narrow Alley, Chengdu, Sichuan
Object: the black brick wall
Inspiration: the feeling of old age

Creative Fabrics Design 1

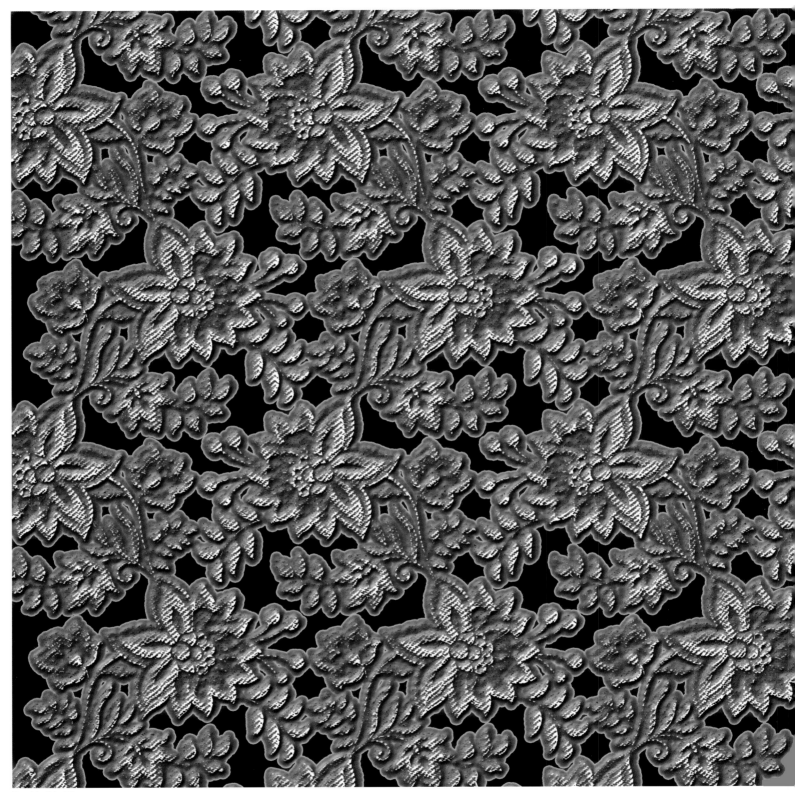

名：追忆花皇（1）
道：天工开物、道法自然
理：石刻、浮雕
法：梭织二方连缀、纬向接回10.8厘米、经向接回64.2厘米
术：梭织提花，印染、轧花、腐蚀
势：2014/2015国际流行趋势之民族文化形态
器：服装面料、装饰布

Name: In Remembrance of the Queen of Flowers (1)
Conviction: heavenly creations, imitation of nature
Principle: stone carving, relief
Method: weaving 2 sides in continuation, zonal tying-in10.8 cm, longitudinal tying-in 64.2 cm
Operation: jacquard weaving, printing and dyeing, embossing,corroding
Trend: 2014/2015 international fashion trend of national culture
Application: clothing fabrics, decorative cloth

时间：2012年7月30日
地点：四川成都宽窄巷子
对象：砖雕、木雕
灵感：凿斑、包浆
Time: July 31, 2012
Location: Broad and Narrow Alley, Chengdu, Sichuan
Object: brick carvings, wood carvings
Inspiration: chisel spot, patina

Creative Fabrics Design 1

名：追忆花皇（2）
道：天工开物，道法自然
理：牙刻、浮雕
法：梭织二方连缀、纬向接回10.8厘米、经向接回64.2厘米
术：梭织提花、印染、轧花、腐蚀
势：2014/2015国际流行趋势之民族文化形态
器：服装面料、装饰布

Name: In Remembrance of the Queen of Flowers (2)
Conviction: heavenly creations, imitation of nature
Principle: ivory carving, relief
Method: weaving 2 sides in continuation, zonal tying-in 10.8 cm, longitudinal tying-in 64.2 cm
Operation: jacquard weaving, printing and dyeing, embossing, corroding
Trend: 2014/2015 international fashion trend of national culture
Application: clothing fabrics, decorative cloth

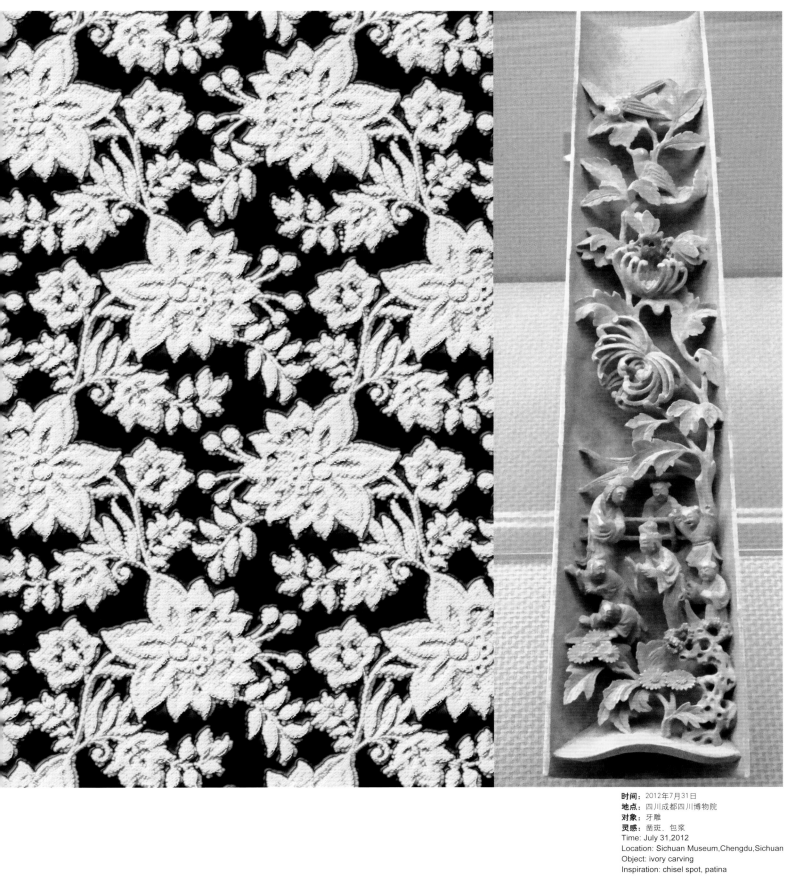

时间：2012年7月31日
地点：四川成都四川博物院
对象：牙雕
灵感：凿斑、包浆
Time: July 31,2012
Location: Sichuan Museum,Chengdu,Sichuan
Object: ivory carving
Inspiration: chisel spot, patina

Creative Fabrics Design 1

名：追忆花皇（3）
道：天工开物、道法自然
理：铜雕；
法：梭织二方连缀、纬向接回10.8厘米、经向接回64.2厘米
术：梭织提花、印染、轧花、腐蚀
势：2014/2015国际流行趋势之民族文化形态
器：服装面料、装饰布

Name: In Remembrance of the Queen of Flowers (3)
Conviction: heavenly creations, imitation of nature
Principle: bronze carving
Method: weaving 2 sides in continuation, zonal tying-in 10.8 cm, longitudinal tying-in 64.2 cm
Operation: jacquard weaving, printing and dyeing, embossing, corroding
Trend: 2014/2015 international fashion trend of national culture
Application: clothing fabrics, decorative cloth

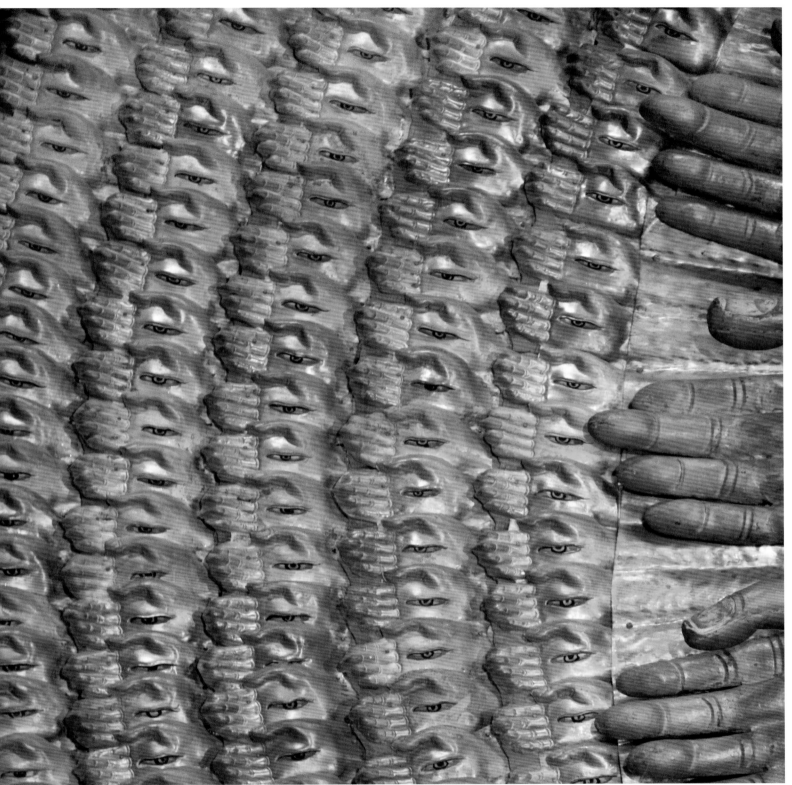

时间：2012年7月30日
地点：川藏塔公寺
对象：铜雕
灵感：金属光感
Time: July 30, 2012
Location: Tagong temple, Sichuan-Tibet
Object: bronze carving
Inspiration: metal glossing

Creative Fabrics Design 1

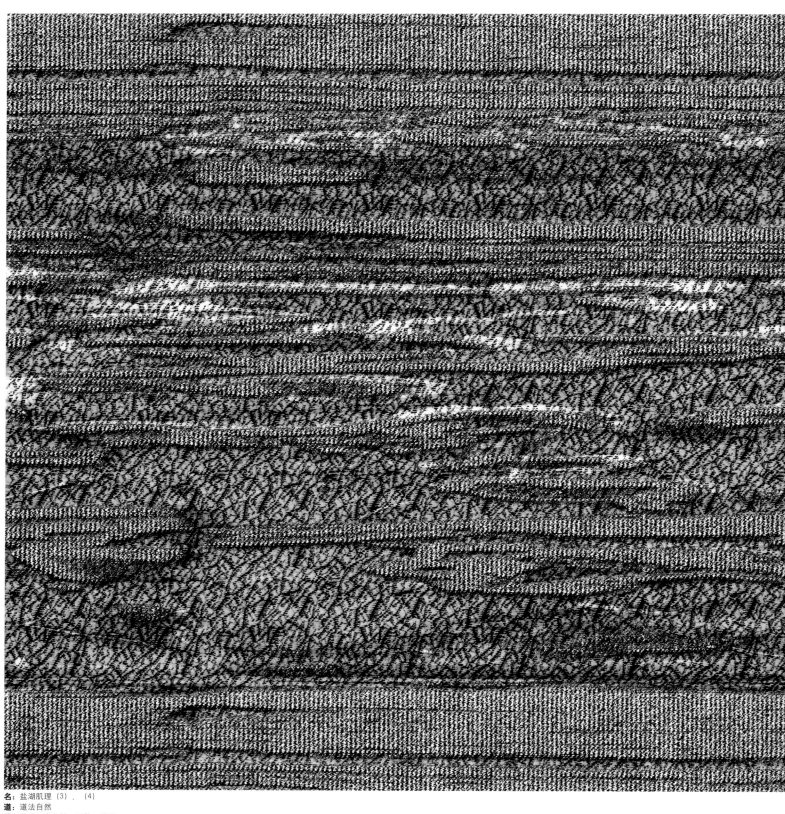

名：盐湖肌理（3）、（4）
道：道法自然
理：风砺、方向性、颗粒、绒质
法：四方连缀、纬向接回35厘米
术：梭织提花、烂花、烧毛、植绒
势：2013/2014国际流行趋势之自然地貌形态
器：服装面料

Name: Salt Lake Texture (3),(4)
Conviction: imitation of nature
Principle: wind effect, directivity, particles, fleece
Method: 4 sides in continuation, zonal tying-in 35 cm
Operation: jacquard weaving, burnt-discharging, singeing, flocking
Trend: 2013/2014 international fashion trend of natural landform
Application: clothing fabrics

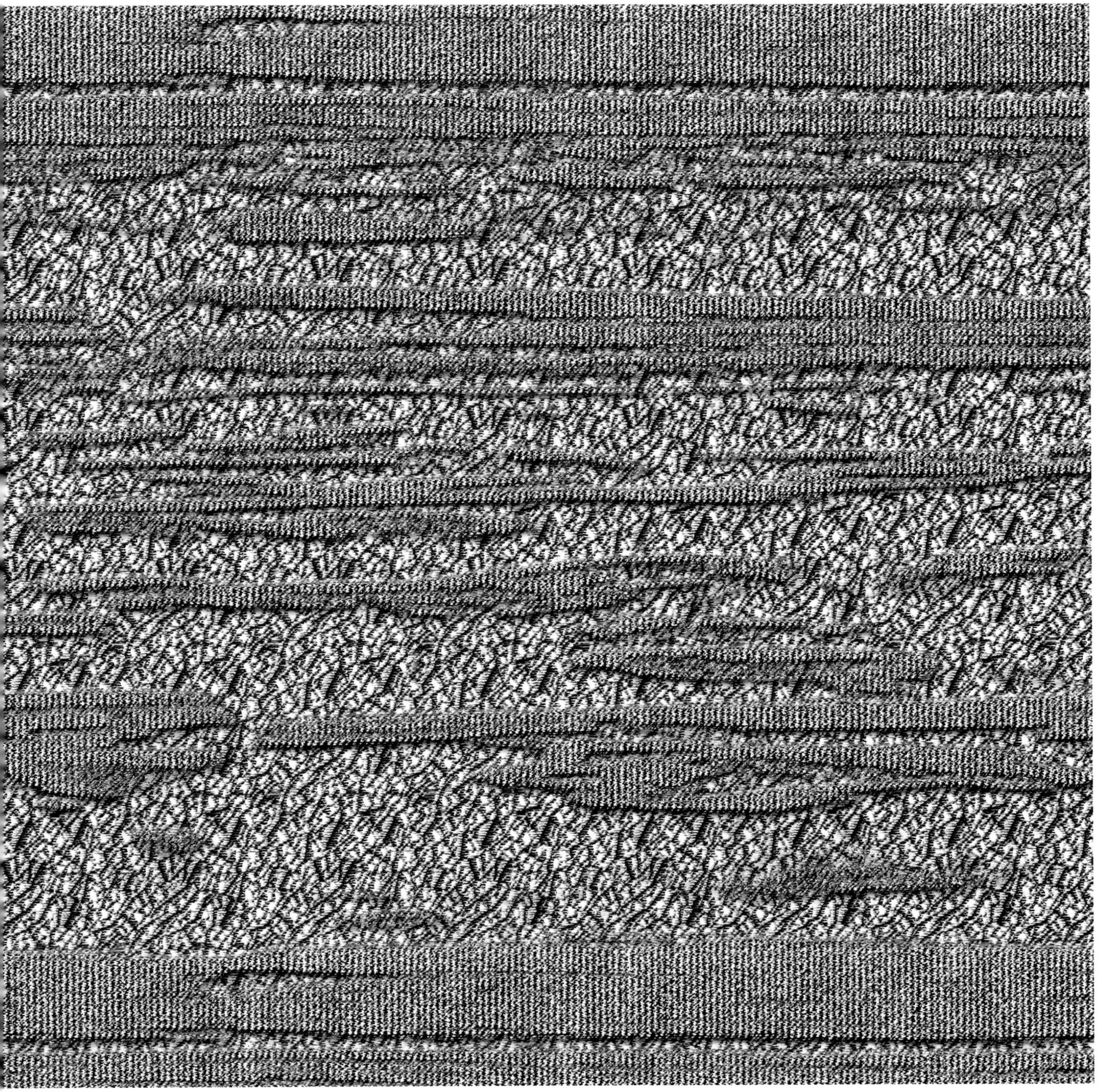

Creative Fabrics Design 1

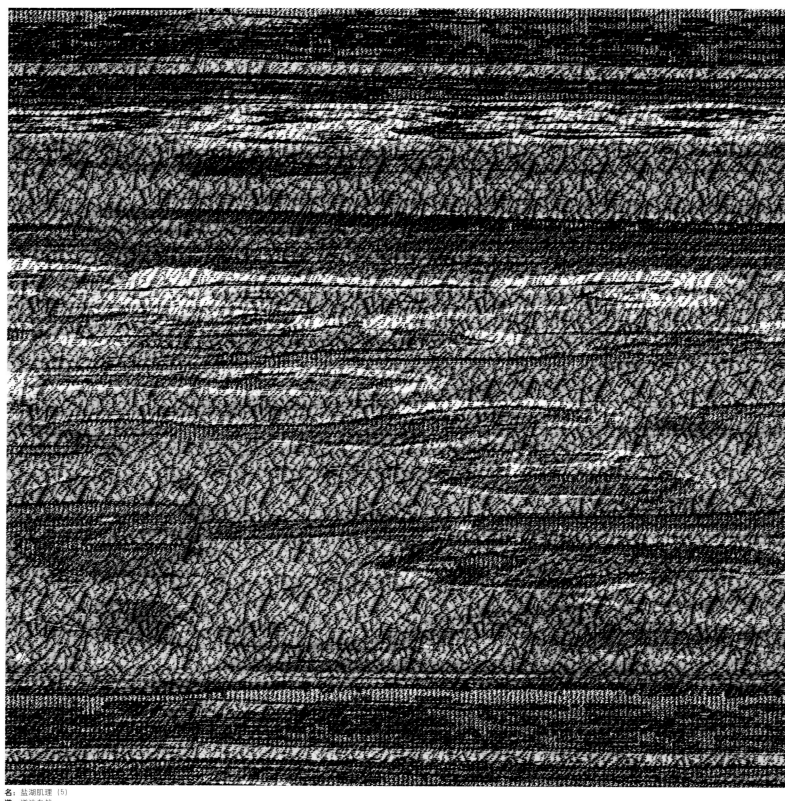

名：盐湖肌理（5）
道：道法自然
理：风砺、方向性、表面颗粒或绒质
法：四方连缀，纬向接回35厘米
术：梭织提花、烂花、烧毛、植绒
势：2013/2014国际流行趋势之自然地貌形态
器：服装面料、装饰布

Name: Saline Lake Texture (5)
Conviction: imitation of nature
Principle: wind effect, directivity, surface particles or fleece
Method: 4 sides in continuation, zonal tying-in 35 cm
Operation: jacquard weaving, burnt-discharging, singeing, flocking
Trend: 2013/2014 international fashion trend of natural landform
Application: clothing fabrics, decorative cloth

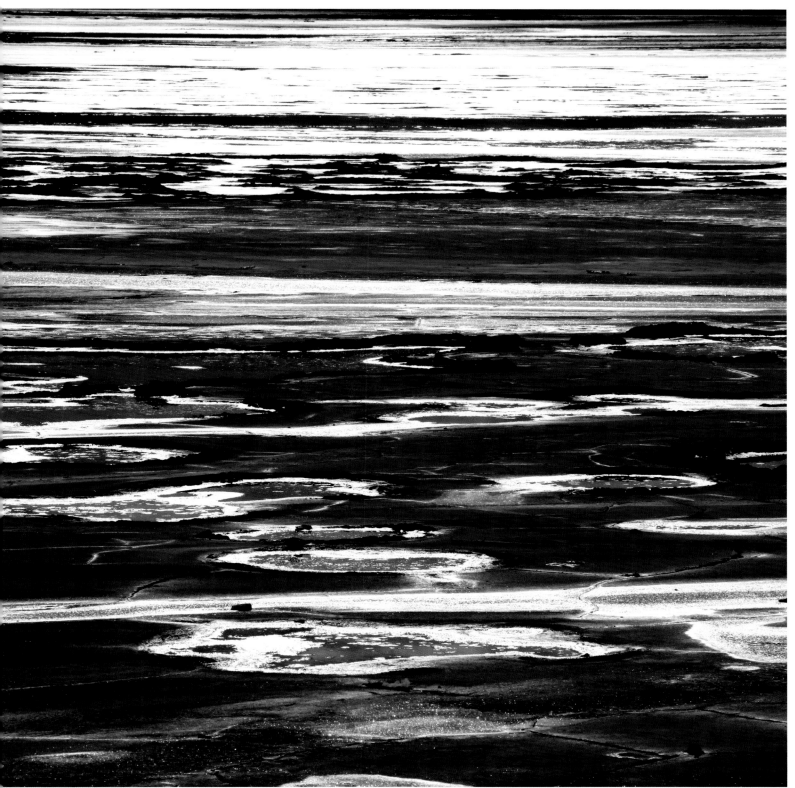

时间：2009年7月19日
地点：青海茶卡
对象：盐湖
灵感：盐在阳光下瞬息万变的光泽
Time: July 19, 2009
Location: Chaka, Qinghai
Object: Salt Lake
Inspiration: gloss, minute to minute variation of salt in the sun

Creative Fabrics Design 1

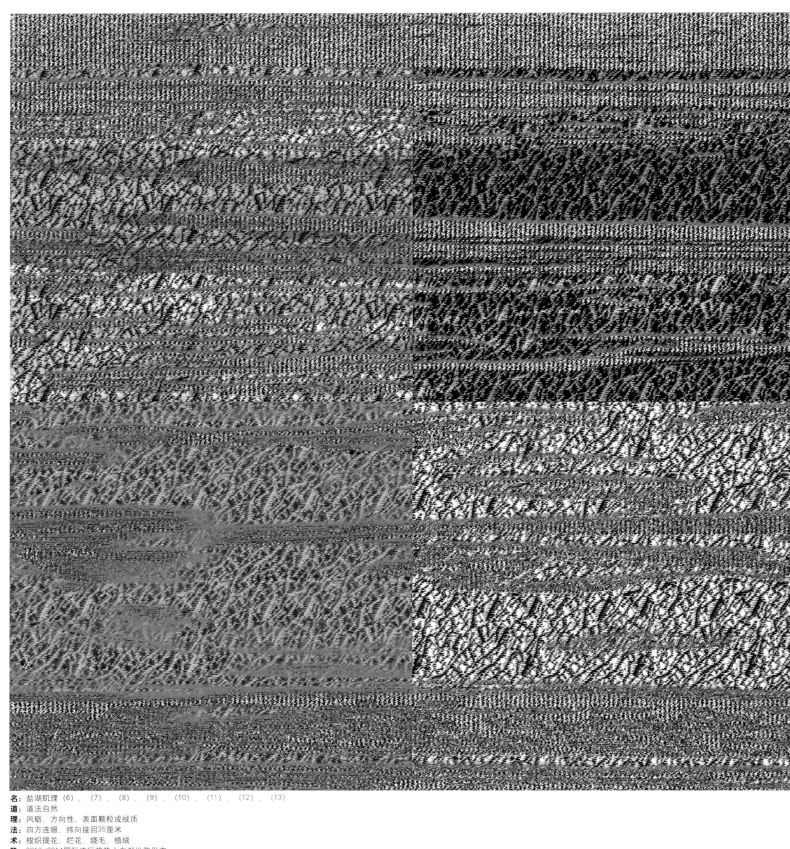

名：盐湖肌理 (6)、(7)、(8)、(9)、(10)、(11)、(12)、(13)
道：道法自然
理：风砺、方向性、表面颗粒或绒质
法：四方连缀、纬向接回35厘米
术：梭织提花、烂花、烧毛、植绒
势：2013/2014国际流行趋势之自然地貌形态
器：服装面料、装饰布

Name: Saline Lake Texture (6), (7), (8), (9), (10), (11), (12), (13)
Conviction: imitation of nature
Principle: wind effect, directivity, surface particles or fleece
Method: 4 sides in continuation, zonal tying-in 35 cm
Operation: jacquard weaving, burnt-discharging, singeing, flocking
Trend: 2013/2014 international fashion trend of natural landform
Application: clothing fabrics, decorative cloth

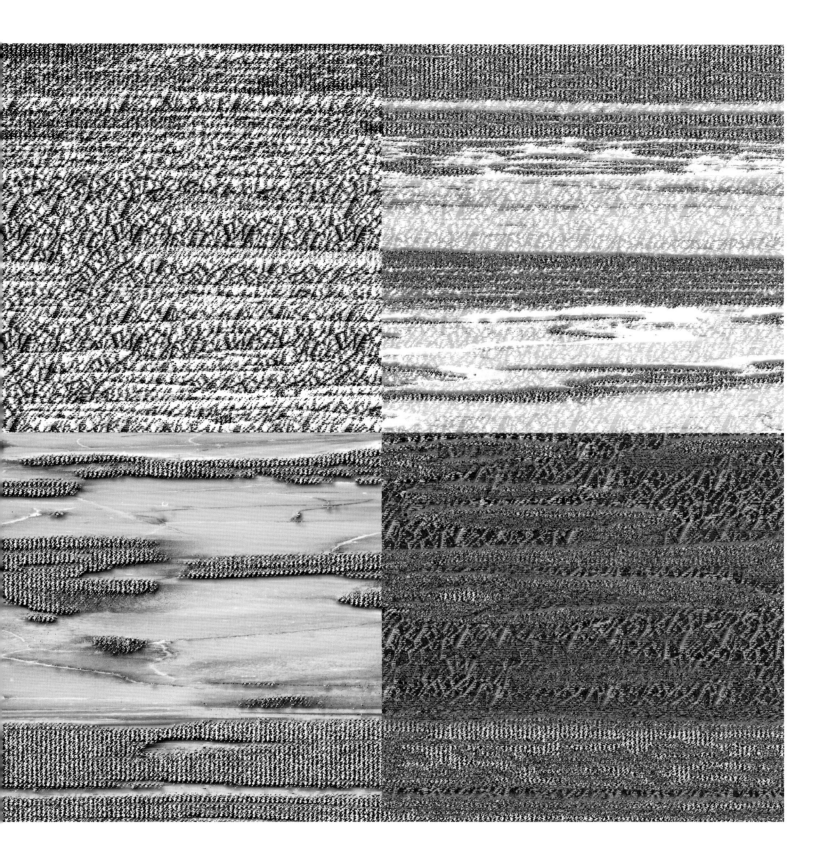

后跋

  本书中的作品是从几千幅中精挑细选而来，大部分的创作时间是去年十月至今年二月底，为新一季时尚趋势的创新作品，可谓快速反应。我一年的纺织品艺术设计作品平均在八九百件，听上去似夸口，但我的学生都知道东华大学第三教学楼北翼的创意面料设计工作室的灯光经常亮至午夜。我的高产已然成为生活的一部分，这种对专业的衷情和热爱令许多人唏嘘不已，然而这份责任和理想说来漫长。

  家父是20世纪五六十年代工艺美术类三大专业之一的染织专业毕业的名校高材生。等到我考学家父就让我冲着染织方向，以承其学、其业、其志。先考苏州丝绸工学院在上海考生中夺得第二名，后考上海纺织高等专科学校落得十二名。家母左思右想不放心上海户口，我便进了纺专。20世纪80年代中期服装业开始兴起，选专业时就一直关注服装设计专业。按照志愿名次本应梦想事成，但我的名字却出现在染织专业里，不过也随了家父的心愿。等我毕业时，纺织业已经每况愈下，而服装业则呈现欣欣向荣的局面。作为夕阳工业的纺锭与织机此时已化作落后的象征。一时间，毕业的学生改行，学校的专业进行调整，行业中的厂企关停并转。我留校任教的上海纺专则被并入中国纺织大学，而纺大不知何故也在这一时期耐不住改名为东华大学，我未进的苏丝院也已升级为苏州大学。盛极一时的纺织产业在上海的凋敝伴随着的是国内曾经具有国际有影响力的研发设计的凋零，一大批有经验、有能力的管理、技术人员和产业工人流失，此种低迷现象延续十多年之久。

  现如今各地大搞创意产业，人们逐渐意识到纺织业也可是创意产业。

  当下中国纺织品艺术设计研发逐渐意识到产学研各方的衔接。"产"要问为谁而产？为什么产？产什么？怎么产？关乎什么？谁求甚解？与我合作的企业在看到我作品的同时，也会听到我这样的询问和自语。

  "学"则必须从象牙塔自以为是，教学上的纸上谈兵中挣脱出来。设计教学更多的是创新思维、技能的表现和传达，英雄不问出处，要尊重一切人才。教学中笔者对学生的专业作业要求有三个方向：一是教学方向，要符合教学要求，解决专业的基础问题，诸如，思维方式、表现技法、艺术理论、工作方式等；二是学术方向，作业最好能参加大赛，放到行业中，让专家评委去检阅，这样评价面就多和宽了；三是商业方向，设计是应该可以用来销售的，我们学的是商业设计，而不是纸上谈兵的所谓"艺术设计"。把作业放到市场上去，理论联系实际。应该建立起这样的概念：中国创造是可以付诸于实施的。

  "研"则需要有思想主导研发，潜心研究市场、探究科学与艺术的结合、主动关心企业所想、工艺的可实现、商业的可销售、时代的可反应。中国创造的建立首先是设计教育在教学中坚持批判思维，扎根民族传统。我们通过各种媒体看国外的灵感来源，那国外设计灵感来自哪里？不能一味被动跟从模仿国际品牌、研发机构的产品和思想。我们应该极力倡导并身体力行，读万卷书、走万里路。我们的设计师应该既走出去从全球文化中吸取灵感，又一定要到中国原住民及原生态文化地区考察，探究中国思想下的符合国际潮流的原创设计，使我们的设计不以牺牲民族文化传承为代价。

  实现美丽中国定须中国创造，其中必需创意。创意可理解为创造性意念，也可作创造性思维活动和创造性活动过程解释。

  我以为纺织品艺术设计中的创意：

  首先是创异，创造不同，不同意味着价值。纺织品艺术设计应该鼓励异化，追求多元。其次是创一，创造第一、唯一，一马当先地领先，独一无二。再次是创益，创造公益、效益和利益，公正的社会效益和可持续发展经济利益。最后是创义，我们的设计可以使人有信仰，为社会、人类做出贡献。

  创意必然为创艺，创造个性，艺术贵在个性、差异，让我们的生活变得更艺术、更自由。

  创意也必将创疑，创造疑问，疑惑使我们对世界充满好奇，对权威的质疑更能开拓我们的思想，文明的进步离不开我们不断地质疑。

  设计创造可以是在于事事在先、在优，但创意更多是创造适宜，在于因时、因地、因人的适合、适宜，使设计符合时代风尚、经济状况及人们的心理预期。

  创意也是创医，设计能够帮助人们摆脱痛苦和烦恼。我们的设计应该让人依重和依靠，贴心创依，好的设计能够主导人们的生活。

  我们的纺织品艺术设计应该帮助人们实现美好理想，创造人们心目中的女神，实现创伊。创意也将创逸，创造安逸。人们向往安逸的生活，面料要给人舒适和有益的影响。

  创意将实现创易，创造一切，包罗万象。创意的核心是创造的知识、财富的增长，不能以牺牲想象力、丧失理想为代价。

  具备如此信念和理想，中国创造才具意义，美丽中国才具有由表及里的美丽。

想到从小到大父亲大人的慈爱和在专业上的谆谆教诲、母亲大人的宠爱，面对你们的恩泽我在此叩首！

忆及成长道路上无数前辈，仙逝的徐仁元、钱国广、邵甲信等诸位专业老师；为笔者授业解惑和事业激励的诸多恩师，韦康、冯刚、宋建社、李浩然等老师，面对你们的关爱学生在此献礼！

感谢从教以来我亲爱的学生们的求知若渴，你们的憧憬和好学是我专业意气风发、勇往直前的源泉，面对莘莘学子的敬爱我在此有礼！

感谢产业界的前辈、学长和同行们，是你们给予我理论联系实践的机会，让我天马行空的创意通过工艺和市场落实到实处，面对你们的厚爱我在此施礼。

一定要感谢各界各路的朋友们，虽然有的是我师长、学长，有的是我的同学，有的是我学生，但长期的相知、相伴、相随，使我们成为人生道路上的知己友人。如水舞深造合伙人阿萍，如珠峰脚下欢笑的旭红，腾格里沙漠何谭惠的滴水相随，攀登乞力马扎罗时大气的Shelley，塞伦盖蒂草原沈姑娘的中国式微笑，桑给巴尔岛颇费周章的Dnna，喀什勇气再现的美公，马来西亚友人Linda坚定又智慧的支持，无私的日本友人佐佐木，授予我第一位外籍会员彰显胆识、德高望重的日本纺织设计协会理事长小川久先生等等无计其数，你们的一颦一笑给予我莫大的信念、力量和无限灵感，让我获益良多，面对你们的垂爱我在此行礼。

感谢学校外语学院的唐再凤老师与范秀华老师，你们容忍我的想法和要求，面对你们创造性的工作我在此敬礼！

感谢国家纺织产品开发中心李斌红主任的卓见、东华大学副校长刘春红博士的识见、著名中国高级定制服装设计师郭培女士的亲见，你们从产学研层面和角度对我作品的见地，明确了我的工作方向并增添了极大的动力，面对你们的抬爱我在此致敬！

最后，感谢所有批评、教诲和帮助我的人。

沈沉

东华大学第三教学楼北翼创意面料设计工作室

2013年3月23日

Postscript

Works in this book were carefully selected from thousands of my designs, most of which were composed between last October and the end of February this year, dedicating to the recent season of fashion trends. So it may be said to be a quick response. It sounds a little complacent to say that the production of my artistic designs of textile fabric amounts to 800 to 900 pieces annually. But all my students have witnessed the midnight lights in the Creative Fabric Design Laboratory, located in the north wing of the No.3 Lecture Building at Donghua University. The rich yielding has become part of my life, which makes some people marvel at my love and passion for the profession. Such great sense of my responsibility and aspiration can only be accounted for in a long story.

My father graduated as an honour student from a prestigious school, majored in dyeing and weaving, one of the three main important majors in industrial arts during the 1950s to 1960s. He wanted me to follow in his steps and work in the same field when I was preparing for the college entrance examination. My examination score qualified me for Suzhou Institute of Silk Industry with the second highest score among the examinees from Shanghai and then in another examination, it put me twelfth for Shanghai Textile College. I landed in the latter to study, as my mother had worried about the possible loss of my registered residency in Shanghai facing the prospect of my leaving home to study in another city. It was in the mid-1980s, when the garment industry began to enjoy its days, and so I kept an eye on the major Garment Design when we students were choosing our majors. My dream should have come true if our grades had played a major role when students were enrolled into different majors. To my surprise my name turned out to be in the major of Dyeing and Weaving, and in a way I was to live up to my father expectation. On my graduation, the textile industry was going from bad to worse, while the garment industry was thriving. Spindle and loom in the so-called sunset industry were regarded as a symbol of retrogressing. Then there came a period of transformation in this field, when graduates switched to other trades, universities adjusted their majors and teaching plans, and firms in the industry were closed, merged or taken over, or shifted to other businesses. During this period, Shanghai Textile College which I graduated from and later worked with as a teacher was merged into the China Textile University, whose name was later changed to Donghua University though. Suzhou Institute of Silk Industry, the school I once missed upgraded to Soochow University. Along with the depression of the once prosperous textile industry in Shanghai, the R&D in design, though enjoyed a leading position domestically and internationally, was doomed to wither. As a result, the industry lost a large number of experienced and efficient managerial staff, technical personnel, and skilled labour. The recession lasted more than a decade.

Now with all engaging in creative industries, people gradually realize that the textile industry may well be a creative industry.The artistic design of textile fabric in current China is gradually seen the awareness of connection among production, learning, and research. As for "production", such questions should be asked: for whom to produce, why to produce, what to produce, how to produce, and what the production will be related to. How much have we explored such questions? The firms I work with would also hear my inquiries and thinking aloud when looking at my designs.

"Learning" should break itself from the complacency of the ivory tower and empty talks in papers. Design teaching is rather the expression and transmission of innovative thinking and capabilities. Talents should be respected because, as the saying goes, the origin of heroes never matters. My requirement for the assignments for students who take my subject has three dimensions: (1) learning—they should be able to meet the syllabus and grasp the fundamental knowledge of the profession, such as ways of thinking, techniques of expression, theory of art, and approaches to work; (2) academic exploration—it is desirable that their works participate in competitions, so as to be examined by experts from the perspective of the industry, leading to a much wider and more multi-faceted evaluation; and (3) commercialisation—designs should be marketable as what we study is commercial design, rather than the so-called art design, or empty talks in papers. Putting their works under the test of the market can link theory with practice. We should establish such a concept: Creativity of China can be put into practice.

"Research" requires idea-led R&D activities, study the market in depth, explore the combination of science and art, take the concerns of firms into consideration, and consider the feasibility of the processes and sales as well as the possible reactions of the times. The first step in the realisation of Creativity of China should have critical thinking as well as be deeply-rooted in national tradition in the teaching of design. When we see foreign inspirations through various media, we should think about the origin of their design revelation. We cannot blindly imitate foreign brands or the products and ideas of foreign R&D institutions. We should advocate and practice being well-read and well-travelled. Our designers should not only go abroad to draw stimulation from the global culture, but also go to the native and aboriginal ecological culture areas in China, so as to explore the original designs generated from Chinese ideas while staying in line with international trend. Therefore our designs will not be accomplished at the costs of sacrificing national cultural heritage.

Beautifying China requires "Creativity of China", which means originality. Creativity can be understood as creative ideas, or creative thinking and creative activity.

According to my understanding, creativity in the artistic design of textile fabrics should include the following aspects. The first one is the creation of differences. Difference means value. Artistic designs of textile should encourage alienation and pursue pluralism. The second one is the creation of number one: being the first one to create, to create what is unique, and what is in the vanguard, in which originality prioritize. The third one is the creation of benefits, including public benefits, profits, interests, social benefits of justice, and benefits of sustainable economic development. The fourth is the manifestation of righteousness, i.e. expressing justice and putting ideas in the right course of moral integrity. Our designs can give people some assurance or conviction, and contribute to the society and human beings.

Creativity surely involves the creation of artistic works and individuality. The value of art lies in individuality. Differences make our life more artistic and freer.

Creativity also means to doubt, which make us curious about the world. The challenge to authority can broaden our minds, and pushes the progress of our civilization.

Designs and creations lie in precedence and excellence. But creativity is more of creating comforts, being suitable and appropriate in accordance with the variability of time, place and the individual. Design should thus be consistent with the trend of the times, the economic situation, and people's expectations.

Creativity is also therapeutic, and design enables people to get out of miseries and worries. Our works should create dependability as well as intimacy to people's sentiments. Good designs can take a lead in people's life.

Our artistic designs of textile should help people realize their ideals and create their own goddess of beauty in their minds. Creativity will create comfort, leading people to a wonderful world of ease and solace they yearn for.

Creativity will achieve change, and create everything. The core of originality is accumulation of created knowledge and wealth, and so it should not be achieved at the expense of imagination, and ideals, but should make contribution to the society and human development.

Only with this conviction and ideal, can our Creativity of China be granted significance and the dream of beautifying China achieved from the outside to the inside.

I kowtow to my dear parents for their bounties. I have had my father's love and inculcation and my mother's doting ever since I was born.

The book reminds me of countless predecessors and teachers along my path of growing up, including the late Renyuan Xu, Guoguang Qian, and Jiaxin Shao. And I salute my mentors, including Kang Wei, Gang Feng and Jianshe Song, Haoran Li for their instructions, doubt dispelling and professional encouragement.

The book can be traced to my dear students, whose hunger for knowledge, and whose aspiration and motivation have always been the source of my professional thrust. I repay with gratitude for their loving admiration.

The outlining of the book owns to my predecessors, seniors and colleagues in the industry, who have provided me with the opportunity to link theory with practice and to land my unconstrained ideas with the help of process and marketing. My respects go to them for their kindness.

The formation of the book has benefited from all the friends with various backgrounds, some of whom are my teachers, students, or classmates. The long-time understanding, companionship and sometimes the occasion that even a little incident, converts the relationship to that of bosom friends on the road of life. They are Ah Ping, my partner of the SWSZ; Hong Xu who laughed heartedly at the foot of Mount Everest; Tanhui He, who followed me in Tengger Desert; Shelley who showed generosity while climbing Mount Kilimanjaro; Ms Shen who smiled in a typical Chinese way on Salem Ghedi Prairie; Dnna under complicated circumstances on Zanzibar Island; the handsome man who appeared once more in Kashgar; Linda, a friend from Malaysia, who supported me with firmness and wisdom; Sasaki, a selfless Japanese friend; Mr. Ogawa, the prestigious chairman of the Japanese Textile Design Association, who granted me the first foreigner membership of the association, etc. Every look and smile of theirs gives me great faith, strength and unlimited inspiration, which have benefited me greatly. I salute all of them for their friendship.

My thanks also go to Associate Professor Zaifeng Tang and Xiuhua Fan for their tolerance of my reckless ideas and urgent requests. Their creative work is greatly appreciated.

The polishing of the book owns much to the insights of Binhong Li, Director of the National Textile Product Development Center; visions of Doctor Chunhong Liu, vice president of Donghua University; and the encouragements from Pei Guo , the renowned haute couture designer. Their views from the perspectives of production, learning and research have shown me a clearer direction and added great incentives to my work. Here I pay tribute to them for their favour.

All in all, I would like to express my gratefulness and love to all those who have given me criticism, instruction, and support.

Chen Shen
Creative Fabric Design Laboratory,
North Wing of No.3 Lecture Building,
Donghua University
23th March, 2013

本书的编写离不开以下人员的帮助和支持：
张嘉翼、李昀、沈露、张帆、郭嘉、王丰、朱铭飞、陈燕、刘洋、刘忠、蔡宏寿、柴方军、李科、李佑平、崔巍、黄婷、陈晓霞、谢方明、徐晶、冯济慈、陈玮、梁爽、岳依天、刘玄曌、胡秋、孙沁、黄静波、张桦、沈丹红、狄晨光、刘炎婷、徐谈、邵琳颖、马文祺、张爱丽、王昆、曹霞、钟英超、张安琪、匡兰、王秋萍、王银、李倩雯、吴琴艳、王昕、俞霜、蔡晨晨、叶菁、邓子怡、陈梦琦、方洁妮、王婷婷、刘柳、张蘭、杨春樱姿、杨昌毅、王、许佳璇、顾莹莹、肖雪怡、徐琳钧、钱文菊、傅鹏瑾、陈如霞、何瑜璠、李粤湘、戴昕、肖静雯、熊黎、卫燕妮、文旭红、钱煜岑、钟其橙、武艺斐、陈玉敏、赵赛群、王艺璇、曲益娇、徐立豪、朱子美、姜安娜、杨俊、曹可、江舟、周蔼蕾、张荣、余苑、陈楚、杨少锋、董健睿、吴顺婷、王凯、陈曼、张云、田宇、鹿崇然、王婧婧、王佩瑶、吴琼、赵园园、刘慧泉、李嫆、沈赞